Springer Tracts in Natural Philosophy

Volume 26

V. G. Romanov

Integral Geometry and Inverse Problems for Hyperbolic Equations

With 21 Figures

Springer-Verlag Berlin Heidelberg New York 1974

V. G. Romanov
Computer Center of the Academy of Science
Novosibirsk, USSR

Title of the Russian Original Edition:
Nekotorye obratnye zadachi dlja uravnenij giperbolicheskogo tipa
Edited by M. M. Lavrent'ev
Publisher: Nauka, Novosibirsk/USSR 1969

AMS Subject Classifications (1970)
35 L 10, 35 R 30, 53 C 65, 86 A 15

ISBN-13: 978-3-642-80783-1 e-ISBN-13: 978-3-642-80781-7
DOI: 10.1007/978-3-642-80781-7

Table of Contents

Introduction

There are currently many practical situations in which one wishes to determine the coefficients in an ordinary or partial differential equation from known functionals of its solution. These are often called "inverse problems of mathematical physics" and may be contrasted with problems in which an equation is given and one looks for its solution under initial and boundary conditions. Although inverse problems are often ill-posed in the classical sense, their practical importance is such that they may be considered among the pressing problems of current mathematical research.

A. N. Tihonov showed [82], [83] that there is a broad class of inverse problems for which a particular non-classical definition of well-posedness is appropriate. This new definition requires that a solution be unique in a class of solutions belonging to a given subset M of a function space. The existence of a solution in this set is assumed a priori for some set of data. The classical requirement of continuous dependence of the solution on the data is retained but it is interpreted differently. It is required that solutions depend continuously only on that data which does not take the solutions out of M.

Let us list some of the most important inverse problems which have been considered. Among these first of all is the seismological inverse kinematic problem, whose formulation goes back to the beginning of this century. Its physical setting is as follows: In a region $\mathscr{D}$ bounded by a surface S, waves are generated by sources at points of the surface S and their travel-times are recorded at various points of the boundary of $\mathscr{D}$; using this information, find the propagation velocities of the disturbances inside $\mathscr{D}$. The problem so stated originated in geophysics. In 1905–1907, Herglotz and Wiechert solved the inverse kinematic problem for a spherical Earth model [34], [84] under the assumption that the propagation velocity increases monotonically with depth. For the first time, this made it possible to make inferences about the Earth's internal structure from seismological data. Geophysicists subsequently made repeated use of the Herglotz-Wiechert method to construct radial models for the velocity distribution of seismic waves inside the Earth. When the propagation velocity is not monotone, the inverse kinematic problem does not have a unique solution. The nature of this lack of uniqueness was studied by M. L. Gerver and V. M. Markuševič [28]. Various results concerning the study of the inverse kinematic problem are contained in [4], [6], [7] and [29].

Another inverse problem which bears on geophysics arises in potential theory. In its general setting, it amounts to the following: Exterior to a region $\mathcal{D}$ bounded by a surface S, one is given the body mass potential, single-layer potential or magnetostatic potential induced by a body of certain density lying inside S. It is required to determine the position, shape and density of the body. The first uniqueness theorems for the inverse problem for a Newtonian potential are due to P. S. Novikov [59]. Later, the inverse problem of potential theory was considered by A. N. Tihonov [82], L. N. Sretenskiĭ [79]–[81]. V. K. Ivanov [36]–[38], M. M. Lavrent'ev [45], [46] and I. M. Rapoport [61], [62]; these authors made essential theoretical advances in the inverse potential problem. A number of new interesting theoretical results on inverse problems for generalized potentials were obtained recently by A. I. Prilepko (see [60]).

Certain questions in quantum mechanics have led to the following problem: Given the spectrum of a second-order differential operator, find the operator. This problem is known as the inverse Sturm-Liouville problem. The first uniqueness theorems were obtained (in the regular case of a finite interval) by V. A. Ambarcumjan [5] and Borg [13]. Later V. A. Marčenko [54] showed that the general inverse Sturm-Liouville problem has to be posed somewhat differently since a single spectrum is insufficient to determine the differential operator. In V. A. Marčenko's setting, the inverse Sturm-Liouville problem reads as follows: Given the spectral function of a second-order differential operator, find the operator. The inverse Sturm-Liouville problem in this form was solved completely by I. M. Gel'fand and B. M. Levitan [26] with a clarification of necessary and sufficient conditions for the spectral function. M.G. Krein studied a slightly different version of the same problem [40]–[44]. A number of results related to the inverse Sturm-Liouville problem have been obtained by V. A. Marčenko [55], [56], L. D. Faddeev [21], [22], L. A. Čudov [19], [20] and others. A detailed discussion of the inverse Sturm-Liouville problem is presented in the books of Z. S. Agranovič and V. A. Marčenko [1] and B. M. Levitan [52].

In the field of partial differential equations, the first uniqueness result for the spectral inverse problem for Schrödinger's equation is due to Ju. M. Berezanskiĭ [9]. Later, M. M. Lavrent'ev considered two classes of inverse problems for partial differential equations in n-dimensional space [47], [48]. A. S. Alekseev showed [3] that certain versions of the inverse problem for second-order partial differential equations whose coefficients depend on just a single variable reduce to the inverse Sturm-

Liouville problem. A little later, a similar procedure was used by A. S. Blagoveščenskiĭ to investigate an inverse problem for systems of equations in elasticity theory [11]. A series of new results on inverse problems for differential equations were obtained in [14]–[16], [49], [50], [57] and [78].

This work deals with various formulations of the inverse problem for hyperbolic equations of mathematical physics. The study of these problems turns out to be closely related to a variety of problems in integral geometry (in the terminology introduced by I. M. Gel'fand [24]). An integral-geometric problem is any problem involving the determination of a function from its integral given over a family of curves or surfaces. An analysis of such problems with applications to group representations is contained in the basic work of I. M. Gel'fand, M. I. Graev and N. Ja. Vilenkin [25]. Some separate problems of integral geometry were examined in [10], [18] and [39]. Generally speaking, integral-geometric problems are ill-posed in the classical sense. The nature of this lack of well-posedness can be seen by the following example considered in Courant's book [18]. Let $u(x, y)$, $x=(x_1, \dots, x_n)$, be a continuous function of x and y which is even in y and let $v(x, y, r)$ be its mean value over a sphere of radius r with center at the point (x, y). Then the problem of determining $u(x, y)$ from its integrals over arbitrary spheres whose centers lie in the hyperplane $y=0$ is equivalent to the problem of finding the solution of the Darboux equation

$$v_{rr}+\frac{n-1}{r}v_r-\Delta_{x,y}v=0 \tag{1}$$

subject to the initial conditions

$$v(x, 0, r)=\psi(x, r), \qquad v_y(x, 0, r)=0. \tag{2}$$

Here $\Delta_{x,y}$ is the Laplacian in x and y and $\psi(x, r)$ is a given function. Problem (1), (2) is Cauchy's problem for a hyperbolic equation with respect to a space variable. It is easy to show that small changes in the function $\psi(x, r)$ can result in finite changes in $v(x, y, r)$. Thus problem (1), (2) is not well-posed in the classical sense. Courant has shown that a solution to the integral-geometric problem is unique in the class of continuous functions $u(x, y)$, where to determine a unique function $u(x, y)$ it suffices to know $\psi(x, r)$ in the cylinder $\{|x|\leq\varepsilon, 0\leq r<\infty\}$, ε being any positive number. Hence, Courant's problem is well-posed according to A. N. Tihonov in any compactum belonging to the space C.

Since the various versions of the inverse problem have led to integral-geometric problems never considered before and since they are of interest in their own right, the author has decided to assign a separate chapter to these integral-geometric problems. Sec. 1 considers the problem of determining a function from its integrals over a family of ellipsoids with one focus fixed and the second situated at a point of a subset of a fixed hyperplane passing through the fixed focus. A uniqueness theorem is proved. In Sec. 2, a generalization of the problem is considered and in Sec. 3, the class of data for which there exists a solution to the problem is characterized. Sec. 4 is devoted to integral-geometric problems in n-dimensional space for a family of curves invariant to parallel translation along a fixed plane. Sufficiency conditions are given for the family of curves of a geometric and analytic nature that assure uniqueness in the problem. Sec. 5 considers a version of the problem involving a finite number of unknown functions. Mathematically, it is equivalent to the determination of the solution of a system of integral equations of the first kind. In Sec. 6, the results of Sec. 4 are carried over to the integral-geometric problem for a family of curves invariant to rotation about a fixed point, and in Sec. 7, to the problem for surfaces invariant to parallel displacement. Sec. 8 occupies a somewhat special place in Chap. I. Considered there is the question: Under what conditions on the coefficients of a second-order hyperbolic linear differential equation does the integral-geometric problem along the extremals induced by the equation have a unique solution? This latter problem has important application in the subsequent investigation of inverse problems.

Chap. II is devoted to an analysis of various versions of the inverse problem for second-order hyperbolic linear partial differential equations. The introductory Sec. 1 presents S. L. Sobolev's method for solving Cauchy's problem for hyperbolic equations, which he developed in the 1930's [72]–[77]. Sec. 2 deals with the one-dimensional inverse problem for the telegraph equation in three-dimensional space using the least amount of information about the solution to the mixed problem for the equation. An existence and uniqueness theorem is proved for the problem. In Sec. 3, two linearizations of the inverse problem for the telegraph equation are considered. One of them reduces to Courant's problem mentioned above and the second to the problem presented in Sec. 1 of Chap. I. Sec. 4 deals with the problem of determining the coefficients of the first derivative terms and the function itself in a second-order partial differential equation knowing the coefficients of the higher-order derivatives. The original problem is reduced to an integral-geometric problem and the uniqueness of a solution is investigated for various cases. In Sec. 5, the linearized n-dimensional inverse kinematic problem is reduced to an integral-geometric problem and a uniqueness theorem is proved. Sec. 6 considers the wave equation in anisotropic media and it is ascertained what information may be obtained about its coefficients from a version of the inverse kinematic problem. The uniqueness theorem obtained there implies, in particular, the result of Herglotz-Wiechert mentioned above. Sec. 7 studies the linearized inverse kinematic problem for the wave equation in anisotropic media. It is converted into an integral-geometric problem similarly to Sec. 5 and a special case is considered. All of the uniqueness theorems obtained in this chapter are constructive in that they all furnish algorithms for determining the required solution.

Chap. III is concerned with the application of the linearized inverse kinematic problem to the study of the Earth's structure. Up to now, almost all studies of the Earth's internal structure have been based on spherically symmetric models. The linearized inverse kinematic problem considered in Sec. 5 of Chap. II makes it possible to construct non-one-dimensional Earth models if one assumes that the propagation velocity of disturbances inside the Earth fluctuates very little relative to geographic coordinates. In Sec. 1, the physical basis for studying the Earth's structure and the question of applicability of the linearized approach to constructing non-spherically symmetric Earth models are discussed. Sec. 2 presents an algorithm for solving the linearized inverse kinematic problem which was programmed at the Computing Center of the Siberian Branch of the USSR Academy of Sciences. Sec. 3 discusses several numerical experiments carried out on the computer. The results of this chapter are due to M. M. Lavrent'ev, A. S. Alekseev, R. G. Muhometov and the author.

The author wishes to take the opportunity to express his sincere gratitude to M. M. Lavrent'ev for his constant encouragement and to A. S. Alekseev for his steady interest and valuable advice.

Chapter I

Some Problems in Integral Geometry

1. Problem of Finding a Function from its Integrals over Ellipsoids of Revolution

Consider the following problem in $(n+1)$-dimensional space $(n \geqslant 1)$: The integrals of a function $u(x, y) = u(x_1, \ldots, x_n, y)$ are known over a family of ellipsoids of revolution with one focus fixed at the origin and the other running over a point set of the hyperplane $y=0$. Determine the function $u(x, y)$ from the known integrals.

Let us give a precise statement of the problem. Let $(x^0, 0) = (x_1^0, \ldots, x_n^0, 0)$ denote the coordinates of the variable focus and $S(x^0, t)$ the surface of the ellipsoid of revolution:[1]

$$r(x, y, 0, 0) + r(x, y, x^0, 0) = t, \tag{1}$$

where $r(x, y, 0, 0)$, $r(x, y, x^0, 0)$ are the distances from the point (x, y) to the points $(0, 0)$, $(x^0, 0)$ respectively. The problem to be considered is to determine the function $u(x, y)$ in terms of the function

$$v(x^0, t) = \iint\limits_{S(x^0, t)} u(x, y)\, d\omega. \tag{2}$$

Here $d\omega$ is the element of solid angle in x, y space with vertex at the origin.

It will be assumed that the function $u(x, y)$ is well-defined for any x, y and belongs to the function space C. From (2) one can see that it is meaningful to pose the question of uniqueness of a solution to this equation only in the class of the even functions in y and henceforth this will be taken into account. In addition we shall assume $u(0, 0) = 0$. This can evidently be done without loss of generality because the value of the

[1] Formulas are numbered consecutively only within each section. In references to formulas within the same chapter the number of the section is added on the left. For example, (4.12) denotes formula (12), Sec. 4. When a formula from another chapter is referred to the number of the chapter is also indicated.

function $u(x, y)$ at the origin can be easily found from (2) by letting the surface $S(x^0, t)$ shrink to the point.

The following uniqueness theorem holds for (2).

Theorem 1.1. *If equation* (2) *has a solution belonging to the function space C which satisfies a Hölder condition in a neighbourhood of the origin, then it is unique.*

By solution here, we mean a function, even in y, vanishing at the origin and satisfying (2).

To make our discussion simpler and more transparent, we shall prove the theorem in two-dimensional space, i.e. for $n=1$, and then we shall show how it is carried over to the case of arbitrary n. In the two-dimensional case, ellipsoids become ellipses and the solid angle ω coincides with the polar angle.

The idea behind the proof of the theorem is to find the moments of the function $u(x, y)$. To this end, it is convenient to pass to polar coordinates r, ω related to Cartesian coordinates for $n=1$ by the formulas

$$x=r\cos\omega,\qquad y=r\sin\omega. \tag{3}$$

The polar equation of an ellipse is given by

$$r=p\,(1-\varepsilon\cos\omega)^{-1}, \tag{4}$$

where p and ε are parameters characterizing the polar distance and eccentricity of the ellipse and are expressed in terms of x^0, t by the formulas

$$\varepsilon=\frac{x^0}{t},\qquad p=\frac{t}{2}(1-\varepsilon^2).$$

Equation (2) becomes

$$\int_0^{2\pi} u(r\cos\omega,\, r\sin\omega)\,d\omega=v(p,\varepsilon), \tag{5}$$

where r is determined by formula (4).

We apply to both sides of equality (5) the operator L defined by the identity

$$\mathrm{L}v\equiv p\frac{\partial}{\partial\varepsilon}\int_0^p v(z,\varepsilon)\frac{dz}{z}. \tag{6}$$

That it is valid to apply L when (5) has a solution belonging to the space C and what the result of its application is, are both substantiated by the following sequence of equalities:

$$\mathrm{L}v = p\frac{\partial}{\partial\varepsilon}\int_0^z \frac{dz}{z}\int_0^{2\pi} u(r_z\cos\omega, r_z\sin\omega)\,d\omega$$

$$= p\frac{\partial}{\partial\varepsilon}\int_0^{2\pi} d\omega\int_0^p u(r_z\cos\omega, r_z\sin\omega)\frac{dz}{z}$$

$$= p\int_0^{2\pi} d\omega\frac{\partial}{\partial\varepsilon}\int_0^{r_p} u(r\cos\omega, r\sin\omega)\frac{dr}{r}$$

$$= \int_0^{2\pi} u(r_p\cos\omega, r_p\sin\omega)\, r_p\cos\omega\,d\omega = \int_{S(p,\varepsilon)} u(x,y)\,x\,d\omega.$$

The subscripts p and z on r introduced here indicate which of these parameters should be substituted in formula (4) in place of the parameter p when computing r. $S(p,\varepsilon)$ denotes the ellipse with parameters p and ε. Note that when we pass in the above equalities from the variable of integration z to the variable r in the inner integral we make use of the relation

$$\frac{dr}{r} = \frac{dp}{p},$$

easily proved with the help of (4).

Applying operator L once more to the resultant relation,

$$\int_{S(p,\varepsilon)} u(x,y)\,x\,d\omega = \mathrm{L}v \tag{7}$$

and letting L^k denote the k-th power of L, we obtain in a similar way the set of relations,

$$\int_{S(p,\varepsilon)} u(x,y)\,x^k\,d\omega = \mathrm{L}^k v \qquad (k=1,2,3,\ldots). \tag{8}$$

Setting $\mathrm{L}^0 v \equiv v(p,\varepsilon)$, we can regard formula (8) as valid for $k=0$ also.

Thus we have uniquely constructed a system of moments on each ellipse. Since $u(x,y)$ is even in y, it is uniquely determined by these moments [2]. In other words, if a solution to (5) exists, it is unique. In addition, the equalities (8) can be used to construct the desired solution.

We shall come back to this question again in Sec. 3 in connection with the existence of a solution to (5).

Consider now the case of arbitrary n. Let us see how the results obtained for ellipses can be carried over to the case of ellipsoids of revolution. Thus, consider (2). Introduce in xy space, along with the fixed Cartesian coordinates, a nonstationary Cartesian system zy, $z=(z_1,\dots,z_n)$, so that the z_1 axis passes through the focus $(x^0, 0)$. Let the variable point (x, y) correspond in the new coordinates to (z, y) and $(x^0, 0)$ to the point $(z^0, 0)=(z_1^0, 0,\dots, 0, 0)$. The nonstationary system of coordinates z, y results from the stationary system x, y by rotation about the origin in the hyperplane $y=0$. Let us denote by Q the matrix of the corresponding orthogonal transformation of the variables z, y into the variables x, y. It is dependent on the direction cosines $q_1,\dots, q_n$ of the radius vector of the point $(x^0, 0)$. Introduce, in addition, spherical coordinates at the point (z, y) by the formulas

$$z_i=r\xi_i \qquad (i=1, 2,\dots, n), \qquad y=r\xi_{n+1}, \tag{9}$$

where ξ_i $(i=1, 2,\dots, n+1)$ are the direction cosines of the radius vector to (z, y) with respect to the nonstationary system of coordinates and r is the distance from the origin to (z, y). Then the equation of an ellipsoid of revolution can be written as

$$r=p(1-\varepsilon\xi_1)^{-1}, \tag{10}$$

where the parameters p, ε are given by

$$\varepsilon=\frac{z_1^0}{t}, \qquad p=\frac{t}{2}(1-\varepsilon^2).$$

Equation (2) can then be rewritten as

$$\iint\limits_{S(q,p,\varepsilon)} u(r\,Q\,\xi)\,d\omega=v(q, p, \varepsilon). \tag{11}$$

Here $\xi=(\xi_1,\dots,\xi_{n+1})$, $q=(q_1,\dots, q_n)$ and $S(q, p, \varepsilon)$ is the surface of the ellipsoid of revolution with parameters q, p, ε. Applying the operators L^k to (11) holding q fixed, we obtain in a similar way to case $n=1$:

$$\iint\limits_{S(q,p,\varepsilon)} u(r\,Q\,\xi)\,z_1^k\,d\omega=\mathrm{L}^k v \qquad (k=0, 1, 2,\dots). \tag{12}$$

Since the transformation from the variables x, y to the variables z, y is orthogonal, the following equality holds:

$$z_1 z_1^0=x_1x_1^0+x_2x_2^0+\cdots+x_nx_n^0,$$

or, stated differently,

$$z_1 = x_1 q_1 + x_2 q_2 + \cdots + x_n q_n, \tag{13}$$

where $q_1, q_2, \ldots, q_n$ are the direction cosines of the radius vector to the variable focus with respect to the $x_1, x_2, \ldots, x_n$ axes. Substituting the expression for z_1 from (13) into (12) and letting the parameter ε approach zero, we easily find, making use of the arbitrariness of the vector $q_1, \ldots, q_n$, the following moments of the function $u(x, y)$:

$$\iint_{S(t)} u(x, y)\, x_1^{\lambda_1}\, x_2^{\lambda_2} \ldots x_n^{\lambda_n}\, d\omega$$

$$(\lambda_i = 0, 1, 2, \ldots),$$
$$(i = 1, 2, \ldots, n).$$

Here $S(t)$ is a sphere of radius $t/2$ with center at the origin. It is evident that our even function of y is uniquely determined by these moments. Hence the validity of the above theorem is proved.

2. Generalization to the Case of Analytic Curves

The method used in the preceding section to find a function from its integrals can easily be carried over (see [65]) to the case of more general surfaces of revolution than ellipsoids. For simplicity of presentation, we shall confine our considerations to the case $n=1$, i.e., to plane curves which are symmetric about the line $y=0$. However, as before, all of our reasoning easily carries over to the case of $(n+1)$-dimensional surfaces obtained by rotating the plane curve about its axis of symmetry.

Consider a family of curves depending on two parameters p and ε $(0 \leqq p < \infty, 0 \leqq \varepsilon \leqq \varepsilon_0)$. Let the polar equation of the curve $S(p, \varepsilon)$ be given by

$$r = p f(\varepsilon, \varepsilon \cos \omega), \tag{1}$$

where $f(\varepsilon, \eta)$ is an analytic function of ε and η in an arbitrarily small, but fixed, neighbourhood of the origin, such that $f(0, 0) \neq 0$ and

$$\frac{\partial}{\partial \eta} f(0, 0) \neq 0.$$

Let the integrals

$$\int_{S(p, \varepsilon)} \varrho(\varepsilon, \varepsilon \cos \omega)\, u(r \cos \omega, r \sin \omega)\, d\omega = v(p, \varepsilon) \tag{2}$$

of $u(x, y)$ be known, where $\varrho(\varepsilon, \eta)$ is an analytic function of ε and η in the same neighbourhood of the origin as $f(\varepsilon, \eta)$ such that $\varrho(0, 0) \neq 0$.

From (2) it is apparent that one can pose the question of uniquely determining only an even function $u(x, y)$ of y from the function $v(p, \varepsilon)$. Therefore throughout the following we shall mean by a solution to (2) an even function of y just as in Sec. 1.

The following theorem is valid.

Theorem 1.2. *If equation* (2) *has a bounded solution over the whole plane belonging to space C and satisfying a Hölder condition in a neighbourhood of the origin, then it is unique.*

The scheme of proof of the theorem is similar to that of the previous theorem. We find the moments of the function on each circle $r=p$ from which the uniqueness of a solution of (2) will follow. In all of our subsequent arguments we shall assume ε to be so small that $\varrho(\varepsilon, \varepsilon \cos\omega)$, $f(\varepsilon, \varepsilon\cos\omega)$ and $\ln f(\varepsilon, \varepsilon\cos\omega)$ are analytic functions of their arguments. Thus these functions can be expanded in Taylor series with respect to their arguments for $0 \leqslant \varepsilon \leqslant \delta$ (δ an arbitrarily small positive number).

Expanding the function $\varrho(\varepsilon, \eta)$ in a series in the variable η and substituting it in formula (2), we obtain

$$v(p, \varepsilon) = \sum_{k=0}^{\infty} \varepsilon^k a_k(\varepsilon)\, v_k(p, \varepsilon), \tag{3}$$

where

$$v_k(p, \varepsilon) = \int\limits_{S(p, \varepsilon)} u(r\cos\omega, r\sin\omega) \cos^k\omega \, d\omega \qquad (k=0, 1, 2, \ldots), \tag{4}$$

$$a_k(\varepsilon) = \frac{1}{k!}\left[\frac{\partial^k}{\partial\eta^k}\varrho(\varepsilon, \eta)\right]_{\eta=0} \qquad (k=0, 1, 2, \ldots). \tag{5}$$

From (4) and the boundedness of the function $u(x, y)$ follows the uniform boundedness of the sequence $v_k(p, \varepsilon)$, $k=0, 1, 2, \ldots$, in the region $\mathscr{D}$ $\{0 \leqslant \varepsilon \leqslant \delta, 0 \leqslant p < \infty\}$. Note that for $\varepsilon=0$, the curves $S(p, \varepsilon)$ coincide with the circles $r=p$; hence, if we can uniquely determine $v_k(p, 0)$ in terms of $v(p, \varepsilon)$, this will prove that $u(x, y)$ is uniquely determined. Let us show that $v_k(p, 0)$ $(k=0, 1, 2, \ldots)$ can actually be found and, moreover, uniquely.

Letting the parameter ε approach zero in (3), we find

$$v_0(p, 0) = \frac{v(p, 0)}{a_0(0)}.$$

We now show how to find $v_1(p, 0)$. For this purpose introduce the operator L defined by

$$\mathrm{L}v \equiv \frac{\partial}{\partial \varepsilon} \int_0^p v(z, \varepsilon) \frac{dz}{z} \tag{6}$$

and apply it to (4). We obtain

$$\mathrm{L}v_k = \frac{\partial}{\partial \varepsilon} \int_0^p \frac{dz}{z} \int_{S(z,\varepsilon)} u(r\cos\omega, r\sin\omega) \cos^k\omega \, d\omega$$

$$= \frac{\partial}{\partial \varepsilon} \int_{S(p,\varepsilon)} \cos^k\omega \, d\omega \int_0^{pf(\varepsilon, \varepsilon\cos\omega)} u(r\cos\omega, r\sin\omega) \frac{dr}{r}$$

$$= \int_{S(p,\varepsilon)} u(r\cos\omega, r\sin\omega) \cos^k\omega \left[\frac{d}{d\varepsilon} \ln f(\varepsilon, \varepsilon\cos\omega)\right] d\omega .$$

From the above formulas one can see that the result of applying the operator L to the functions $v_k(p, \varepsilon)$ is a new sequence of functions $\mathrm{L}v_k$ ($k=0, 1, 2, \ldots$) also uniformly bounded in the region $\mathscr{D}$. If in the last integral we expand the function in brackets in powers of $\varepsilon\cos\omega$ and we integrate the resulting series term by term, we find

$$\mathrm{L}v_k = b_0(\varepsilon)\, v_k(p, \varepsilon) + \sum_{n=0}^{\infty} b_n(\varepsilon)\, \varepsilon^{n-1} v_{k+n}(p, \varepsilon), \tag{7}$$

where

$$b_0(\varepsilon) = \left[\frac{\partial}{\partial \varepsilon} \ln f(\varepsilon, \eta)\right]_{\eta=0},$$

$$b_n(\varepsilon) = \frac{1}{n!} \left\{\frac{\partial^n}{\partial \eta^n} \left[n \ln f(\varepsilon, \eta) + \varepsilon \frac{\partial}{\partial \varepsilon} \ln f(\varepsilon, \eta)\right]\right\}_{\eta=0} \tag{8}$$

$$(n=1, 2, 3, \ldots).$$

By the conditions imposed upon the function $f(\varepsilon, \eta)$ the coefficient $b_1(0) = [(\partial/\partial\eta)\, f(\varepsilon, \eta)]_{\varepsilon=\eta=0} \neq 0$. Due to this we can now obtain an equation for determining $v_1(p, 0)$. To this end apply the operator L to (3) and in so doing, formally interchange the order of the summation operator and L on the right-hand side. This yields

$$\mathrm{L}v = \sum_{k=0}^{\infty} \frac{\partial}{\partial \varepsilon} (\varepsilon^k a_k(\varepsilon)) \int_0^p v_k(z, \varepsilon) \frac{dz}{z}$$

$$+ \sum_{k=0}^{\infty} \varepsilon^k a_k(\varepsilon)\, \mathrm{L}v_k . \tag{9}$$

The series on the right-hand side of this formula converge uniformly in the region $\mathscr{D}'\{0\leqslant\varepsilon\leqslant\delta, 0\leqslant p\leqslant P\}$ where P is any positive number. Indeed, the sequence $\mathrm{L}v_k$ $(k=0, 1, 2,\ldots)$ is uniformly bounded in the region $\mathscr{D}\{0\leqslant\varepsilon\leqslant\delta, 0\leqslant p<\infty\}$ and the sequence $\int_0^p v_k(z, \varepsilon)\, z^{-1}\, dz$ $(k=0, 1, 2,\ldots)$ is also uniformly bounded in $\mathscr{D}'$. The latter follows from the uniform boundedness of the sequence $v_k(p, \varepsilon)$ $(k=0, 1, 2,\ldots)$ in $\mathscr{D}$ and the fulfillment of a Hölder condition for $u(x, y)$ and hence, of one for the functions $v_k(p, \varepsilon)$ in a neighbourhood of the origin. From the above, the uniform convergence in $\mathscr{D}'$ of the series in (9) follows from the uniform convergence in $\mathscr{D}$ of the series $\sum_{k=0}^{\infty}(\partial/\partial\varepsilon)\,[\varepsilon^k a_k(\varepsilon)]$, $\sum_{k=0}^{\infty}\varepsilon^k a_k(\varepsilon)$. Thus, the required uniform convergence of the series is established. Hence referring to the familiar theorems of analysis, we can affirm the validity of the interchange of the order of L performed above. We now make use of formula (7) for $\mathrm{L}v_k$. Substituting this expression for $\mathrm{L}v_k$ into (9), we get

$$\mathrm{L}v=c_0(\varepsilon)\int_0^p v_0(z, \varepsilon)\frac{dz}{z}+\sum_{k=1}^{\infty}\varepsilon^{k-1}c_k(\varepsilon)\int_0^p v_k(z, \varepsilon)\frac{dz}{z}$$

$$+d_0(\varepsilon)\, v_0(p, \varepsilon)+\sum_{k=1}^{\infty}\varepsilon^{k-1}d_k(\varepsilon)\, v_k(p, \varepsilon). \tag{10}$$

The coefficients c_k and d_k $(k=0, 1, 2,\ldots)$ in the respective series are easily computed in terms of the coefficients a_k and b_k $(k=0, 1, 2,\ldots)$. In this connection $d_1(0)=a_0(0)\, b_1(0)\neq 0$. Letting the parameter ε approach zero in (10) and dividing by $d_1(0)$, we obtain the following Volterra equation for $v_1(p, 0)$:

$$v_1(p, 0)+\lambda\int_0^p v_1(z, 0)\frac{dz}{z}=f_1(p), \tag{11}$$

where λ is a numerical parameter. The solution of (11) is generally speaking not unique since for $\lambda<0$, (11) has an eigenfunction of the form $Cp^{-\lambda}$, where C is an arbitrary constant. The equation has no other eigenfunctions. Note, however, that we must seek a bounded solution of (11). Since for $\lambda<0$ the function $Cp^{-\lambda}$ becomes infinite if $C\neq 0$ and $p\to\infty$ we can assert that a solution to (11) is unique.

Applying the operator L to (3) k times in succession letting ε approach zero and dividing by the numerical coefficient $a_0(0)\, b_1^k(0)\neq 0$, we obtain for the function $w_k(z)=v_k(\exp z, 0)$ the Volterra integral equation

$$w_k(z)+\int_{-\infty}^{z}[\lambda_0(z-\xi)^{k-1}+\lambda_1(z-\xi)^{k-2}+\cdots+\lambda_{k-2}(z-\xi)+\lambda_{k-1}]$$

$$\times\, w_k(\xi)\, d\xi=f_k(z), \qquad (k=1, 2, 3,\ldots),$$

where $\lambda_0, \lambda_1, \ldots, \lambda_{k-1}$ are numerical coefficients and $f_k(z)$ is a continuous function expressible in terms of the functions $[L^m v]_{\varepsilon=0}$ $(m=0, 1, 2, \ldots, k)$. The equation is easily shown by reduction to a differential equation with constant coefficients to have at most k independent eigenfunctions each of which tends to infinity as $z \to +\infty$; hence, a bounded solution of this equation is unique. Thus, all the functions $v_k(p, 0)$ $(k=0, 1, 2, \ldots)$ are uniquely determined. This implies the uniqueness of a solution to (2).

The following example shows that if the condition of the theorem concerning the boundedness of the function $u(x, y)$ could be weakened, it would be insignificant. Even for the case of ellipses considered above, if the solution is not bounded over the whole plane, there are examples in which a weight function may be chosen so that the required function will not be uniquely determined from its integrals.

Indeed, consider the function

$$u(r\cos\omega, r\sin\omega) = r^{\gamma} \sum_{k=1}^{N} A_k \cos k\omega, \tag{12}$$

where γ is any positive number, N is any positive integer, and the A_k are arbitrary numerical coefficients. Let us choose $\varrho(\varepsilon, \eta) = (1-\eta)^{\gamma}$ to be the weight function.

Then, if $S(p, \varepsilon)$ is an ellipse with parameters p, ε given by (1.4), we have

$$\int_{S(p,\varepsilon)} (1-\varepsilon\cos\omega)^{\gamma}\, u(r\cos\omega, r\sin\omega)\, d\omega = 0$$

for any p and $0 \leqslant \varepsilon < 1$. This implies that besides the trivial solution, the homogeneous equation (2) has a non-trivial solution given by (12).

3. Existence Theorem for the Case of Ellipses

We have already said before that integral-geometric problems are ill-posed according to Hadamard and the concept of well-posedness introduced for these problems by A. N. Tihonov appears to be more natural. Accordingly it is to be assumed that the solution of a problem exists on some set of data. Nevertheless it is always of interest to study the the characteristic properties of the class of data for which a solution exists. An investigation of these properties helps us understand how narrow the class of data is and what regularization methods are advantageous to apply in solving the corresponding problem. In this section it will be shown by example of the integral-geometric problem for ellipses considered in Sec. 1, how, having a method of constructing the solution, one can determine necessary and sufficient properties on the class of data for which a solution to the integral-geometric problem will exist.

The technique applied here is easily carried over to the other integral-geometric problems to be considered in the subsequent sections. As might be expected, the set of data for which a solution exists is very narrow, namely, it is not closed in most common function spaces.

Let us consider how the relations in (1.8) can be used to express $u(x, y)$ explicitly in terms of $v(p, \varepsilon)$ and simultaneously to investigate what properties must be ascribed to the set of those functions $v(p, \varepsilon)$ for which a solution to (1.5) exists. In so doing we shall slightly contract the class of functions which we have shown can be uniquely determined from their integrals over ellipses. Namely, we shall consider functions $u(r, \omega)$ satisfying the following conditions:

1°. Each $u(r, \omega)$ is continuous in the disc $0 \leqslant r \leqslant r_0$ and even in ω; and $u(0, \omega) = 0$. Here r_0 is an arbitrary positive number.

2°. In a neighbourhood of the origin of the polar coordinate system, each $u(r, \omega)$ satisfies a Hölder condition

$$|u(r, \omega)| \leqslant A r^{\mu} \qquad (\mu > 0), \tag{1}$$

where A and μ are constants.

3°. Each $u(r, \omega)$ satisfies the inequality

$$\sum_{k=0}^{\infty} \max_{0 \leqslant r \leqslant r_0} |u_k(r)| < \infty, \tag{2}$$

where

$$\begin{aligned} u_0(r) &= \frac{1}{2\pi} \int_{-\pi}^{\pi} u(r, \omega)\, d\omega, \\ u_k(r) &= \frac{1}{\pi} \int_{-\pi}^{\pi} u(r, \omega) \cos k\omega \, d\omega \qquad (k = 1, 2, 3, \ldots). \end{aligned} \tag{3}$$

The class of functions for which conditions 1°–3° are fulfilled will be labeled U.

Let us change slightly the formulation of the problem. Consider in the xy plane a circle of radius r_0 and only those ellipses considered above which lie in this circle. The problem will now be to determine the function $u(r, \omega)$ inside the disc $0 \leqslant r \leqslant r_0$ from its integrals over this family of ellipses.

Let the function $v(p, \varepsilon)$ be such that a solution to (1.5) exists. We let the parameter ε tend to zero in (1.8). As a result, the ellipses go over into circles of radius p and the limiting process yields the equality

$$\int_{-\pi}^{\pi} u(p, \omega) \cos^k \omega \, d\omega = p^{-k} [\mathrm{L}^k v]_{\varepsilon=0} \qquad (k=0, 1, 2, \ldots). \tag{4}$$

On the basis of this equality alone we can construct for fixed r a Fourier series for the function $u(r, \omega)$ which, by condition 3°, converges for $u \in U$. Thus, knowing the integrals over ellipses of eccentricity varying in the interval $0 \leqslant \varepsilon \leqslant \delta$ where δ is an arbitrarily small positive integer, we can determine the function $u(r, \omega)$ in the disc $0 \leqslant r \leqslant r_0$ and, hence, determine the integrals of it over all the ellipses lying within the circle. This implies that by specifying the function $v(p, \varepsilon)$ in an arbitrarily small region $0 \leqslant \varepsilon \leqslant \delta$, we can determine it entirely for $\delta \leqslant \varepsilon \leqslant 1$ also. This in turn implies that a solution to the stated problem does not exist for an arbitrary continuous function $v(p, \varepsilon)$. The result obtained is related to the fact that the formulated problem is not well-posed according to Hadamard. Indeed, no closed linear manifold of functions which belong to the function spaces C^k, H, L_p or W_p^l possesses the property obtained above for the function $v(p, \varepsilon)$. In the following we shall show what necessary and sufficient properties the set of functions $v(p, \varepsilon)$ must have for a function $u(r, \omega)$ satisfying (1.5) to exist.

Consider a family of the linear operators M_k defined by

$$\mathrm{M}_0 v \equiv \frac{1}{2\pi} \mathrm{L}^0 v \equiv \frac{1}{2\pi} v(p, \varepsilon),$$

$$\mathrm{M}_k v \equiv \frac{1}{\pi} \sum_{j=0}^{[k/2]} (-1)^j \left(\sum_{l=0}^{[k/2]-j} C_k^{2(j+1)} C_{j+1}^l \right) p^{2j-k} \mathrm{L}^{k-2j} v, \tag{5}$$

$$(k=1, 2, 3, \ldots).$$

Using the system of relations (1.8), we can easily show that the result of applying the operators M_k to (1.5) can be expressed as

$$\frac{1}{2\pi} \int_{S(p,\varepsilon)} u(r, \omega) \{[t+\sqrt{t^2-1}]^k + [t-\sqrt{t^2-1}]^k\}_{t=\cos\omega/(1-\varepsilon\cos\omega)} \, d\omega = \mathrm{M}_k v$$

$$(k=0, 1, 2, \ldots). \tag{6}$$

In this connection one should keep in mind that for $|t|<1$ the expressions in brackets are complex-valued.

Nevertheless the whole expression in braces is real. For $\varepsilon=0$ equality

(6) can be written as

$$\frac{1}{\pi}\int_{-\pi}^{\pi} u(p,\omega)\cos k\omega\, d\omega = [\mathrm{M}_k v]_{\varepsilon=0} \qquad (k=1,2,3,\ldots). \tag{6'}$$

Hence, if there exists a solution to (1.5) belonging to the set U, it can be expressed in terms of the function $v(p,\varepsilon)$ by

$$u(p,\omega)=\sum_{k=0}^{\infty}[\mathrm{M}_k v]_{\varepsilon=0}\cos k\omega. \tag{7}$$

The convergence of this series for $u\in U$ follows from (6′) and condition 3° on $u(r,\omega)$. We now study what properties the set V of functions $v(p,\varepsilon)$ (the image of set U) must have under the correspondence defined by (1.5).

Theorem 1.3. *The image V of the set U possesses the following properties:*

1°. *The functions $\mathrm{M}_k v$ $(k=0,1,2,\ldots)$ constructed for arbitrary $v\in V$ from formulas* (5), *exist and are continuous, and $[\mathrm{M}_k v]_{p=0}=0$.*

2°. *For any $v\in V$ the series*

$$\sum_{k=0}^{\infty}\max_{0\leqslant p\leqslant r_0}|\mathrm{M}_k v|_{\varepsilon=0}=\alpha_v \tag{8}$$

is convergent.

3°. *The function $u(r,\omega)$ constructed from $v(p,\varepsilon)\in V$ through* (7), *satisfies a Hölder condition*

$$|u(r,\omega)|\leqslant Ar^{\mu} \qquad (\mu>0). \tag{9}$$

4°. *The function $v(p,\varepsilon)$ satisfies the identity*

$$\int_{-\pi}^{\pi}\sum_{k=0}^{\infty}[\mathrm{M}_k v]_{\substack{\varepsilon=0\\ p\to r}}\cos k\omega\, d\omega\equiv v(p,\varepsilon). \tag{10}$$

in which the expression $[\mathrm{M}_k v]_{\substack{\varepsilon=0\\ p\to r}}$ is to be understood as follows: first we compute the function $\mathrm{M}_k v$ at $\varepsilon=0$, and then we replace p by r as given by formula (1.4).

Properties 1°–3° obviously follow from the respective properties of the functions $u(r,\omega)$ and equalities (6), (6′) and (7). To prove identity (10) it is sufficient to note that the function $u(r,\omega)$ determined by (7) must satisfy (1.5). From this identity there follows, in particular, the inequality

$$|v(p,\varepsilon)| \leqslant 2\pi \sum_{k=0}^{\infty} \max_{0\leqslant p\leqslant r_0} |\mathrm{M}_k v|_{\varepsilon=0} = 2\pi\,\alpha_v$$

Theorem 1.4. *For* (1.5) *to have a solution belonging to the set U, it is necessary and sufficient that the function* $v(p,\varepsilon)$ *belong to set V.*

The necessity follows from Theorem 1.3. Let us prove that $v(p,\varepsilon)$ belonging to the set V is also sufficient for the existence of a solution. Indeed, the series

$$\sum_{k=0}^{\infty} [\mathrm{M}_k v]_{\varepsilon=0} \cos k\omega = u(p,\omega) \tag{11}$$

constructed from a given function $v(p,\varepsilon)\in V$, converges uniformly by property 2°, and determines a function $u(p,\omega)$ which is continuous in the variables p and ω. Moreover the Fourier coefficients of $u(p,\omega)$ coincide with $[\mathrm{M}_k v]_{\varepsilon=0}$ $(k=0,1,2,\ldots)$. By virtue of properties 1°–3° of $v(p,\varepsilon)$, the function $u\in U$. It remains to show that u satisfies (1.5). Using $u(p,\omega)$ we construct the function

$$\tilde{v}(p,\omega) = \int_{S(p,\varepsilon)} u(r,\omega)\, d\omega. \tag{12}$$

We shall show that $\tilde{v}(p,\omega)=v(p,\omega)$.

Observe first of all that $\tilde{v}\in V$. Applying M_k to (12), we obtain

$$[\mathrm{M}_k\tilde{v}]_{\varepsilon=0} = [\mathrm{M}_k v]_{\varepsilon=0} \qquad (k=0,1,2,\ldots). \tag{13}$$

Introduce now a new function

$$w(p,\varepsilon) = v(p,\varepsilon) - \tilde{v}(p,\varepsilon). \tag{14}$$

It is evident that the function $w\in V$. At the same time by linearity of the operators M_k, applying (13), we are led to a system of equalities:

$$[\mathrm{M}_k w]_{\varepsilon=0} = 0 \qquad (k=0,1,2,\ldots), \tag{15}$$

which with identity (10), allows us to assert that the function $w(p,\varepsilon)\equiv 0$, i.e. $\tilde{v}(p,\varepsilon)=v(p,\varepsilon)$. This means that each such function provides the solution of (1.5) determined by (10). By Theorem 1.1. it is unique.

4. Determination of a Function from its Integrals over a Family of Curves Invariant to Displacement

In this section we shall consider a family of curves invariant to all motions involving parallel translation along a hyperplane (in n-dimensional space). For brevity the family of curves will be said to be invariant to displacement along the hyperplane.

In the following we shall examine various versions of integral-geometric problems for a family of curves invariant to displacement.

I. Consider a region $\mathscr{D}\{0\leqslant y\leqslant H\}$ of x, y space, $x=(x_1,\ldots, x_n)$, and an $(n+1)$-parameter family of curves in $\mathscr{D}$ invariant to displacement along the hyperplane $y=0$. Let each curve of the family be a smooth arc of finite length in $\mathscr{D}$, with its endpoints in the hyperplane $y=0$. A natural integral-geometric problem is then the following: Given the integrals of a function $u(x, y)$ with respect to arclength along the above family of curves, determine the function $u(x, y)$ in $\mathscr{D}$. It will be shown that if certain natural conditions are imposed on the differentiability of the family of curves, then the problem will have a unique solution in the class of continuous functions having compact support.

We list now the properties of the family of curves to be essentially used in the following. Our scheme of presentation is as follows: first geometric restrictions are imposed on the family from which analytic properties of the family are then derived with the help of appropriate lemmas. The latter are used to establish uniqueness theorems in integral geometry. Let us denote the radius of curvature of a curve at a fixed point by R, the unit principal normal vector by n, and the unit vector along the y axis by β. The scalar product of the vectors β and n will be denoted by (β, n). Let the family of curves satisfy the following conditions:

1°. It is invariant to the displacement along the hyperplane $y=0$.

2°. Each curve of the family is twice continuously differentiable with respect to arclength.

3°. For each point $(\xi, \eta)\in\mathscr{D}$, $\xi=(\xi_1,\ldots, \xi_n)$, there exists a curve $L(\xi, \eta)$, belonging to the family, such that its endpoints lie in the hyperplane $y=0$ and the point (ξ, η) is its vertex (i.e. the point of the curve most distant from the hyperplane $y=0$.)

4°. For the family of curves $L(\xi, \eta)$ there exist positive numbers δ, r_0, R_0, q such that the portion of each curve $L(\xi, \eta)$ enclosed in the region $\mathscr{G}_\eta=\mathscr{D}\cap\mathscr{D}_\eta$, $\mathscr{D}_\eta=\{\eta-y\leqslant\delta\}$, is a continuous curve for all $\eta\in[0, H]$ for which the inequalities

$$0<r_0\leqslant R\leqslant R_0<\infty,\quad -(\beta, n)\geqslant q>0$$

are satisfied.

5°. The family of curves $L(\xi, \eta)$ fills out the region $\mathscr{D}$ smoothly in the sense that the radius of curvature R of each $L(\xi, \eta)$ and (β, n) are continuously differentiable functions of the parameter η.

We now convert the above requirements on the family of curves into analytical form. To this end, we introduce as running coordinate on the curve $L(\xi, \eta)$ the arclength s' measured from the vertex of $L(\xi, \eta)$, regarded positive in one direction from the point (ξ, η) and negative in the opposite direction. Then, on the basis of properties 1°–3°, the equation of $L(\xi, \eta)$ can by represented parametrically as

$$\begin{aligned} x &= \xi + f(s, \eta), \\ y &= \eta - \psi(s, \eta), \end{aligned} \tag{1}$$

where $f(s, \eta)$, $\psi(s, \eta)$ are twice continuously differentiable functions with respect to the first argument such that $\psi(s, \eta) \geqslant 0$, $f(0, \eta) = 0$, $\psi(0, \eta) = 0$, $\psi_s(0, \eta) = 0$. The last condition follows from property 3°. Indeed, for fixed η the function $y = y(s, \eta)$ attains a maximum value at $s = 0$ and, since it is continuously differentiable with respect to s, its partial derivative with respect to s at this point, must equal zero. Let us see what limitations condition 4° imposes on the functions f and ψ. In this connection we shall recall some information from differential geometry. Let $r = r(s)$ be the parametric equation of a curve in n-dimensional space. Here r is the radius vector to a variable point of the curve and s is arclength. Then

$$\frac{dr}{ds} = \sigma, \qquad \frac{d^2 r}{ds^2} = \frac{d\sigma}{ds} = \frac{1}{R} n, \tag{2}$$

wherein σ is the unit tangent vector to the curve at point s, n is the unit principal normal vector on the concave side of the curve and R is the radius of curvature of the curve. Note that the first formula of (2) in conjunction with $\psi_s(0, \eta) = 0$ yields $|f_s(0, \eta)| = 1$. The second formula and condition 4° lead to the following inequality in $\mathscr{G}_\eta$:

$$-(\beta, n) = -R\left(\beta, \frac{d^2 r}{ds^2}\right) = R \frac{\partial^2}{\partial s^2} \psi(s, \eta) \geqslant q > 0,$$

or, taking into account the inequality for the radius of the curvature,

$$0 < \frac{q}{R_0} \leqslant \frac{\partial^2}{\partial s^2} \psi(s, \eta) \leqslant \frac{1}{r_0}. \tag{3}$$

Thus condition 4° implies the uniform boundedness of $(\partial^2/\partial s^2)\, \psi(s, \eta)$ in $\mathscr{G}_\eta$.

Condition 5° implies the continuity of the functions

$$\frac{\partial^3}{\partial s^2\, \partial \eta} \psi(s, \eta), \qquad \frac{\partial^3}{\partial s^2\, \partial \eta} f(s, \eta)$$

in the region $\mathscr{G} = \cup_{0 \leqslant \eta \leqslant H} \mathscr{G}_\eta$.

Below we shall need the following lemma.

Lemma 1. *Let a family of curves satisfy conditions* 1°–5° *in the region*

$\mathscr{D}$. Then in the region $\mathscr{G}_\eta$ the equation of the curve $L(\xi, \eta)$, $(\xi, \eta) \in \mathscr{D}$, can be represented in the form of two single-valued branches,

$$x = \xi + \varphi_j(\sqrt{\eta - y}, \eta), \quad (j = 1, 2), \tag{4}$$

where $\varphi_j(p, \eta)$ $(j = 1, 2)$ are continuous vector functions of their arguments in $\mathscr{G}$ together with the partial derivatives

$$\frac{\partial^{k+l}}{\partial p^k \, \partial \eta^l} \varphi_j(p, \eta), \quad (j = 1, 2), (k = 1, 2; l = 0, 1; k + l \leqslant 2),$$

such that

$$\varphi_j(0, \eta) = 0, \left| \frac{\partial}{\partial p} \varphi_j(p, \eta) \right|_{p=0} = \sqrt{\frac{2}{\psi_{ss}(0, \eta)}} \geqslant \sqrt{2r_0} > 0, \quad (j = 1, 2). \tag{5}$$

Proof. Consider the second equation in (1) and rewrite it as

$$\sqrt{\eta - y} = \sqrt{\psi(s, \eta)}. \tag{*}$$

The function $\psi(s, \eta)$, as we found above, vanishes at $s = 0$ along with its first partial derivative with respect to s. From (3) it follows that for fixed η it has a minimum at that point. Thus in a neighbourhood of the minimum it behaves like a convex function. Hence it follows that to each value of $p = \sqrt{\eta - y} > 0$, the parameter s has exactly two values in the region $\mathscr{G}_\eta$. To the point $p = 0$ corresponds the value of the parameter $s = 0$. Hence, for fixed η, the equation of the curve (*) can be represented by two single-valued branches;

$$s = F_j(p, \eta), \quad (j = 1, 2), \tag{6}$$

such that to positive and negative values of s there correspond different branches and $F_j(0, \eta) = 0$, $j = 1, 2$. Let us now show that the functions $F_j(p, \eta)$, $j = 1, 2$, have continuous first order partial derivatives and continuous second order partials $(\partial^2/\partial p^2)\, F_j(p, \eta)$, $(\partial^2/\partial p \, \partial \eta)\, F_j(p, \eta)$ in $\mathscr{G}$. Indeed, substituting the expressions for s from formula (6) into equality (*) and differentiating the resulting identity with respect to p and η, we find

$$\begin{aligned} 1 &= \left\{ \tfrac{1}{2} [\psi(s, \eta)]^{-1/2} \frac{\partial}{\partial s} \psi(s, \eta) \right\}_{s = F_j(p, \eta)} \frac{\partial}{\partial p} F_j(p, \eta), \\ 0 &= \left\{ \tfrac{1}{2} [\psi(s, \eta)]^{-1/2} \frac{\partial}{\partial s} \psi(s, \eta) \right\}_{s = F_j(p, \eta)} \frac{\partial}{\partial \eta} F_j(p, \eta) \\ &\quad + \left[\tfrac{1}{2} (\phi(s, \eta))^{-1/2} \frac{\partial}{\partial \eta} \psi(s, \eta) \right]_{s = F_j(p, \eta)} \end{aligned} \tag{7}$$

The first equality in (7) shows that the partial derivatives of $F_j(p, \eta)$, $j=1, 2$, with respect to p are continuous in the region $\mathscr{G}$ since convexity of $\psi(s, \eta)$ in the variable s implies that the expression in braces in (7) does not vanish for $p>0$. As $p\to 0$ the partial derivatives each have a finite limit. The result is

$$\frac{\partial}{\partial p} F_j(p, \eta)\bigg|_{p=0} = (-1)^j \sqrt{\frac{2}{\psi_{ss}(0, \eta)}}, \qquad (j=1, 2). \tag{8}$$

In this formula, $j=1$ is the branch of the curve $L(\xi, \eta)$ corresponding to negative values of s and $j=2$ the branch corresponding to positive values of s. The continuity of the partial derivatives $(\partial/\partial\eta)\, F_j(p, \eta), j=1, 2$, is easily proved by use of the second equality in (7). We can check the continuity of the derivatives $(\partial^2/\partial p^2)\, F_j(p, \eta)$, $(\partial^2/\partial\eta\, \partial p)\, F_j(p, \eta)$, $j=1, 2$, in the region $\mathscr{G}$ in exactly the same way. To do this, we merely have to differentiate the first equality in (7) in the variables p, η and make use of the smoothness of the function $\psi(s, \eta)$ mentioned above. The indeterminacies arising as $p\to 0$ are removed in a simple way.

Substituting now the expression for s from (6) into the first formula of (1) and denoting

$$f\,(F_j(p, \eta)) = \varphi_j(p, \eta), \qquad (j=1, 2),$$

we arrive at formulas (4). The differentiability properties of the functions $\varphi_j(p, \eta)$, $j=1, 2$, mentioned in the lemma become obvious from the corresponding properties of the functions $f\,(s, \eta)$ and $F_j(p, \eta)$. The validity of formulas (5) is easily verified. Indeed:

$$\varphi_j(0, \eta) = f\,(F_j(0, \eta), \eta) = f\,(0, \eta) = 0,$$

$$\left|\frac{\partial}{\partial p}\varphi_j(p, \eta)\right|_{p=0} = \{|\, f_s(s, \eta)|_{s=F_j(p,\eta)} \cdot |F_{jp}(p, \eta)|\}_{p=0}$$

$$= \left|f_s(0, \eta)\right| \cdot \left|\frac{\partial}{\partial p} F_j(p, \eta)\right|_{p=0} = \sqrt{\frac{2}{\phi_{ss}(0, \eta)}} \geqslant \sqrt{2r_0} > 0.$$

The last inequality follows from formula (3). Thus the lemma is proved.

II. We return now to the integral-geometric problem formulated above. Let the integrals

$$\int_{L(\xi, \eta)} u(x, y)\, ds = v(\xi, \eta), \qquad (\xi, \eta) \in \mathscr{D} \tag{9}$$

of the function $u(x, y)$ be known over the family of curves $L(\xi, \eta)$.

The following theorem is valid.

Theorem 1.5. *Let $u(x, y)$ be a continuous function with compact support* *in the region $\mathscr{D}$ and let the family of curves $L(\xi, \eta)$ satisfy conditions* 1°–5°. *Then the function $u(x, y)$ is uniquely determined by the function $v(\xi, \eta)$.*

Lemma 1 states that each curve $L(\xi, \eta)$ can be represented in the form (4) in a neighbourhood of its vertex. Assume at first that this representation is valid not only in a neighbourhood of the vertex of the curve $L(\xi, \eta)$ but in general the equation of $L(\xi, \eta)$ can be represented everywhere in the region $\mathscr{D}$ in the form (4) (i.e. $\delta = H$ in condition 4°). Let us prove, on this assumption, the validity of the above theorem.

$$v(\xi, \eta) = \sum_{j=1}^{2} \int_0^{\eta} \left[u(\xi + \varphi_j(p, \eta), y) \sqrt{1 + \left| \frac{\partial}{\partial p} \varphi_j(p, \eta) \right|^2 \left(\frac{\partial p}{\partial y} \right)^2} \right]_{p=\sqrt{\eta - y}} \cdot dy. \tag{10}$$

Taking the Fourier transform of both sides of the equation in the variable ξ and denoting

$$v_\lambda(\eta) = \int_{-\infty}^{\infty} v(\xi, \eta)\, e^{i(\lambda, \xi)}\, d\xi,$$

where the symbol (λ, ξ) stands for the scalar product of the vectors λ, ξ, we find

$$v_\lambda(\eta) = \sum_{j=1}^{2} \int_{-\infty}^{\infty} e^{i(\lambda, \xi)}\, d\xi \int_0^{\eta} \Bigg[u(\xi + \varphi_j(p, \eta), y)$$

$$\cdot \sqrt{1 + \frac{1}{4} \left| \frac{\partial}{\partial p} \varphi_j(p, \eta) \right|^2 \frac{1}{p^2}} \Bigg]_{p=\sqrt{\eta - y}} dy$$

$$= \sum_{j=1}^{2} \int_0^{\eta} \Bigg\{ \frac{1}{p} \sqrt{p^2 + \frac{1}{4} \left| \frac{\partial}{\partial p} \varphi_j(p, \eta) \right|^2}$$

$$\cdot \int_{-\infty}^{\infty} u(\xi + \varphi_j(p, \eta), y)\, e^{i(\lambda, \xi)}\, d\xi \Bigg\}_{p=\sqrt{\eta - y}} \cdot dy$$

$$= \sum_{j=1}^{2} \int_0^{\eta} \Bigg\{ \frac{1}{p} \sqrt{p^2 + \frac{1}{4} \left| \frac{\partial}{\partial p} \varphi_j(p, \eta) \right|^2}\, e^{-i(\lambda, \varphi_j(p, \eta))}$$

$$\cdot \int_{-\infty}^{\infty} u(x, y)\, e^{i(\lambda, x)}\, dx \Bigg\}_{p=\sqrt{\eta - y}} dy$$

$$=\int_0^\eta u_\lambda(y)\, K_\lambda(\sqrt{\eta-y}, y)\,(\eta-y)^{-1/2}\, dy.$$

Here $u_\lambda(y)$ denotes the Fourier transform of the function $u(x, y)$ and the function $K_\lambda(p, \eta)$ is given by the formula

$$K_\lambda(p, \eta)=\sum_{j=1}^{2}\sqrt{p^2+\frac{1}{4}\left|\frac{\partial}{\partial p}\varphi_j(p, \eta)\right|^2}\; e^{-i(\lambda, \varphi_j(p, \eta))}. \tag{11}$$

The resulting equation

$$v_\lambda(\eta)=\int_0^\eta u_\lambda(y)\, K_\lambda(\sqrt{\eta-y}, \eta)\,(\eta-y)^{-1/2}\, dy, \quad (0\leqslant\eta\leqslant H) \tag{12}$$

is a Volterra integral equation of the first kind. From formula (11) and the properties of the functions $\varphi_j(p, \eta), j=1, 2$, it follows that the function $K_\lambda(p, \eta)$ occurring in the kernel of (12) is continuous in its arguments along with the first order partial derivatives. Besides

$$K_\lambda(0, \eta)=\tfrac{1}{2}\sum_{j=1}^{2}\left|\frac{\partial}{\partial p}\varphi_j(p, \eta)\right|_{p=0}=\sqrt{\frac{2}{\psi_{ss}(0, \eta)}}\geqslant\sqrt{2r_0}>0. \tag{13}$$

Making use of these properties, we can easily transform (12) into a Volterra equation of the second kind (see, for example, [58]). Indeed, applying to (12) the operator L:

$$\mathrm{L}v_\lambda\equiv\frac{1}{\pi}\frac{d}{dt}\int_0^t\frac{v_\lambda(\eta)}{\sqrt{t-\eta}}\, d\eta, \tag{14}$$

we obtain

$$\mathrm{L}v_\lambda=\frac{1}{\pi}\frac{d}{dt}\int_0^t\frac{d\eta}{\sqrt{t-\eta}}\int_0^\eta u_\lambda(y)\,\frac{K_\lambda(\sqrt{\eta-y}, \eta)}{\sqrt{\eta-y}}\, dy$$

$$=\frac{1}{\pi}\frac{d}{dt}\int_0^t u_\lambda(y)\left\{\int_y^t\frac{K_\lambda(\sqrt{\eta-y}, \eta)}{\sqrt{(t-\eta)(\eta-y)}}\, d\eta\right\} dy$$

$$=u_\lambda(t)\, K_\lambda(0, t)+\int_c^t u_\lambda(y)\left\{\frac{1}{\pi}\frac{\partial}{\partial t}\int_y^t\frac{K_\lambda(\sqrt{\eta-y}, \eta)}{\sqrt{(t-\eta)(\eta-y)}}\, d\eta\right\} dy.$$

Hence, Eq. (12) reduces to

$$u_\lambda(t)+\int_0^t u_\lambda(y)\, R_\lambda(y,t)\, dy=\frac{1}{K_\lambda(0,t)}\, \mathrm{L}v_\lambda, \qquad (0\leqslant t\leqslant H) \tag{15}$$

in which the kernel $R_\lambda(y, t)$ is given by the formula

$$R_\lambda(y,t)=\frac{1}{\pi K_\lambda(0,t)}\frac{\partial}{\partial t}\int_y^t \frac{K_\lambda(\sqrt{\eta-y},\eta)}{\sqrt{(t-\eta)(\eta-y)}}\, d\eta\,. \tag{16}$$

If the integration variable η is replaced by the new variable

$$\eta=t\cos^2\Theta+y\sin^2\Theta$$

the expression for the kernel $R_\lambda(y, t)$ becomes

$$R_\lambda(y,t)=\frac{2}{\pi K_\lambda(0,t)}\frac{\partial}{\partial t}\int_0^{\pi/2} K_\lambda(\sqrt{t-y}\cos\Theta,\, t\cos^2\Theta+y\sin^2\Theta)\, d\Theta$$

$$=\frac{2}{\pi K_\lambda(0,t)}\int_0^{\pi/2}\left\{\frac{\partial}{\partial p}K_\lambda(p,\eta)\frac{\cos\Theta}{2\sqrt{t-y}}\right.$$

$$\left.+\frac{\partial}{\partial\eta}K_\lambda(p,\eta)\cos^2\Theta\right\}_{\substack{p=\sqrt{t-y}\cos\Theta,\\ \eta=t\cos^2\Theta+y\sin^2\Theta}} d\Theta\,.$$

It is seen from the last formula that the kernel of (15), generally speaking, has a weak polar singularity along the diagonal. But function

$$\sqrt{t-y}\cdot R_\lambda(y,t)$$

remains continuous and therefore by iteration of the kernel, Eq. (15) can be reduced to an equation with continuous kernel [58]. The latter can be solved by the method of successive approximations.

From the above it follows that the function $u_\lambda(y)$ is uniquely determined by the function $v_\lambda(\eta)$. Hence the function $u(x, y)$ also is uniquely determined by $v(\xi, \eta)$.

In order to write, if only formally, the inversion formula for Eq. (9), it is necessary to impose more stringent conditions upon the function $u(x, y)$ than compact support and continuity. It suffices, for example, to require also the fulfillment of Dirichlet conditions with respect to $u(x, y)$.

We show now how the result obtained above can be used to prove Theorem 1.5 in the general case. Decompose the region $\mathscr{D}$ by means of the hyperplanes $y=\text{const.}$ into a finite number of regions $\mathscr{D}_i\{y_i \leqslant y \leqslant y_{i+1}\}$, $i=1, 2, \ldots, N$, $(y_1=0)$, so that $\max_{1\leqslant i\leqslant N}(y_{i+1}-y_i)\leqslant\delta$. Consider now the region $\mathscr{D}_1$.

Because of Lemma 1 in the region $\mathscr{D}_1$ the equation of any curve $L(\xi, \eta)$ with vertex belonging to $\mathscr{D}_1$ can be expressed in the form (4). Therefore the function $u(x, y)$ from the foregoing is uniquely determined in $\mathscr{D}_1$ by $v(\zeta, \eta)$, with $(\zeta, \eta)\in\mathscr{D}_1$. Let us show that $u(x, y)$ is uniquely determined in the region $\mathscr{D}_2$ also. For this purpose consider curves $L(\xi, \eta)$ for which $(\xi, \eta)\in\mathscr{D}_2$. Each of the curves $L(\xi, \eta)$ passes through $\mathscr{D}_1$ and $\mathscr{D}_2$. But in $\mathscr{D}_1$ the value of $u(x, y)$ can be regarded now as known. Therefore splitting the integral of $u(x, y)$ along each curve $L(\xi, \eta)$ into two, one over the part of the curve $L(\xi, \eta)$ passing through $\mathscr{D}_2$ and the other over the part of the curve passing through $\mathscr{D}_1$, we arrive at the following problem: Find the function $u(x, y)$ in $\mathscr{D}_2$ from its integrals over the portion of the curves $L(\xi, \eta)$, $(\xi, \eta)\in\mathscr{D}_2$, lying in the region $\mathscr{D}_2$. Since for any curve $L(\xi, \eta)$, the inequality $\eta-y\leqslant y_2-y_1\leqslant\delta$ is satisfied in $\mathscr{D}_2$, the part of any curve $L(\xi, \eta)$, $(\xi, \eta)\in\mathscr{D}_2$, enclosed in $\mathscr{D}_2$ can be represented by Eq. (4) because of Lemma 1. Thus, we are again led to a problem already investigated from which follows the unique determination of the function $u(x, y)$ in $\mathscr{D}_2$ in terms of the function $v(\xi, \eta)$, $(\xi, \eta)\in\mathscr{D}_2$. Repeating this recursive process for the regions $\mathscr{D}_3, \mathscr{D}_4, \ldots, \mathscr{D}_N$ we can show that the function $u(x, y)$ is uniquely determined in terms of the function $v(\xi, \eta)$ in the whole region $\mathscr{D}$.

III. The above method of investigating the integral-geometric problem allows certain generalizations in two directions. First, there is the extension of the class of curves for which the uniqueness theorem holds and, second, there is the generalization of the integral-geometric problem itself. We begin by looking at the first. Analyzing the proof of Theorem 1.5, we easily see that the uniqueness theorem remains valid for curves having an order of tangency at the vertex which is different from that considered above. We shall not be concerned here with the geometric aspects of the problem. We merely note that it is possible to consider families of curves for which representation (4) holds with the replacement of $\sqrt{\eta-y}$ in this formula by the more general expression $(\eta-y)^\alpha$ $(0<\alpha<1)$ and replacement of (5) by

$$\varphi_j(0, \eta)=0, \left|\frac{\partial}{\partial p}\varphi_j(p, \eta)\right|_{p=0}\geqslant\varepsilon>0, \quad (j=1, 2). \tag{5'}$$

In this case the proof of Theorem 1.5 remains valid with the operator L replaced by a more general operator in conformity with the theory of integral equations having a singularity such as in the generalized Abel equation [58].

We generalize now the formulation of the integral-geometric problem. Namely, we assume that we are given the integrals of $u(x, y)$ with weight function $\varrho(x-\xi, y, \eta)$, i.e., to each curve $L(\xi, \eta)$, $(\xi, \eta)\in\mathscr{D}$, of the family there corresponds the functional

$$\int_{L(\xi,\eta)} \varrho(x-\xi, y, \eta)\, u(x, y)\, ds = v(\xi, \eta), \tag{17}$$

and it is required to find $u(x, y)$ from these functionals. On the basis of the method used to prove Theorem 1.4 it is easy to prove the following theorem.

Theorem 1.6. *Let $u(x, y)$ be a continuous function of compact support in the region $\mathscr{D}$, let the family of curves $L(\xi, \eta)$ satisfy conditions* 1°–5° *and let the weight function $\varrho(x-\xi, y, \eta)$ be continuously differentiable with respect to all arguments for any $(x, y)\in\mathscr{D}$, $(\xi, \eta)\in\mathscr{D}$ and such that $\varrho(0, \eta, \eta)\neq 0$, $\eta\in[0, H]$. Then the function $u(x, y)$ is uniquely determined by its integrals* (17).

To convince ourselves of this, we shall look at the scheme of proof of Theorem 1.5 and indicate those changes in the formulas which need to be made in this case. It will be noted that $y\equiv\eta-p^2$, where $p=\sqrt{\eta-y}$. Denote

$$\varrho(\varphi_j(p, \eta), \eta-p^2, \eta)=\varrho_j(p, \eta), \quad (j=1, 2). \tag{18}$$

It is evident that the functions $\varrho_j(p, \eta)$, $j=1, 2$, have continuous first order partial derivatives in the variables p, η. Making transformations in (17) analogous to those of Theorem 1.5 we arrive at (12) in which the function $K_\lambda(p, \eta)$ in this case has the form

$$K_\lambda(p, \eta)=\sum_{j=1}^{2} \varrho_j(p, \eta) \sqrt{p^2+\tfrac{1}{4}\left|\frac{\partial}{\partial p}\varphi_j(p, \eta)\right|^2}\; e^{-i(\lambda, \varphi_j(p, \eta))} \tag{19}$$

Moreover,

$$K_\lambda(0, \eta)=\tfrac{1}{2}\varrho(0, \eta, \eta)\sum_{j=1}^{2}\left|\frac{\partial}{\partial p}\varphi_j(p, \eta)\right|_{p=0}\neq 0 \quad (0\leqslant\eta\leqslant H). \tag{20}$$

Thus the properties of the new function $K_\lambda(p, \eta)$, determined by formula (20), are totally analogous to those of the previous function $K_\lambda(p, \eta)$. Therefore all further transformations of (12) and the results obtained are entirely valid in this case too. Thus the validity of the theorem is established.

5. The Integral-Geometric Problem for *m* Functions

I. In the previous section we considered the problem of determining one function from its integrals over a family of curves. Mathematically, this problem was reduced to investigating an integral equation of the first kind. Consider now the integral-geometric problem in the case where the unknown function $u(x, y)$ is an m-dimensional vector, i.e. $u=(u_1, \ldots, u_m)$. Suppose we have r families of curves $L_k(\xi, \eta)$, $k=1, 2, \ldots, r$, and suppose each of these families generates m_k scalar equations for determining the function $u(x, y)$. In this case it is natural to assume that in such equations each component of the function $u(x, y)$ has its own weight function as coefficient. So, consider the following system of equations:

$$\int_{L_k(\xi, \eta)} (v_{kl}, u)\, ds = v_{kl}(\xi, \eta)$$

$$\begin{matrix} k=1, 2, \ldots, r, \\ l=1, 2, \ldots, m_k, \end{matrix} \qquad (m_1+m_2+\cdots+m_r=m), \tag{1}$$

where (v_{kl}, u) is the scalar product of the vectors v_{kl} and u and ds is the element of arclength along the curve $L_k(\xi, \eta)$.

For the system of equations (1) the following theorem is valid.

Theorem 1.7. *Let the vector function $u(x, y)$ have compact support and be continuous in the region $\mathscr{D}\{0 \leqslant y \leqslant H\}$. Let the families of curves $L_k(\xi, \eta)$, $k=1, 2, \ldots, r$, satisfy conditions* 1°–5° *of Sec.* 4 *and let the weight functions $v_{kl} \equiv v_{kl}(x-\xi, y, \eta)$, $k=1, 2, \ldots, r$; $l=1, 2, \ldots, m_k$, be continuous along with their first order partial derivatives and be such that the system of vectors $v_{kl}(0, \eta, \eta)$, $k=1, 2, \ldots, r$; $l=1, 2, \ldots, m_k$, $\eta \in [0, H]$, is linearly independent. Then the function $u(x, y)$ is uniquely determined by the system of equalities* (1).

To prove the theorem we use the method of Sec. 4. According to Lemma 1 the equation of each curve $L_k(\xi, \eta)$ can be represented by an equation of the form (4.4). The functions $\varphi_j(\sqrt{\eta - y}, \eta)$ contained in this equation for the k families of curves $L_k(\xi, \eta)$ will be denoted by $\varphi_{jk}(\sqrt{\eta - y}, \eta)$. Introduce also the notation

$$v_{kl}(\varphi_{jk}(p, \eta), \eta - p^2, \eta) = v_{jkl}(p, \eta). \tag{2}$$
$$(j=1, 2; \quad k=1, 2, \ldots, r; \quad l=1, 2, \ldots, m_k)$$

Then take Fourier transforms with respect to the variable $\xi=(\xi_1, \ldots, \xi_n)$

in system (1). We arrive just as in Sec. 4 at the system of equations

$$\int_0^\eta (u_\lambda(y), K_{\lambda kl}(\sqrt{\eta-y}, \eta))\,(\eta-y)^{-1/2}\,dy = v_{\lambda kl}(\eta),\qquad k=1,2,\dots,r;\quad l=1,2,\dots,m_k, \tag{3}$$

where $u_\lambda(y)$ and $v_{\lambda kl}(\eta)$ are the Fourier transforms of $u(x, y)$ and $v_{kl}(\xi, \eta)$ and the $K_{\lambda kl}(p, \eta)$ are m-dimensional vector functions determined by the formulas (cf. formula (4.19))

$$K_{\lambda kl}(p, \eta) = \sum_{j=1}^{2} v_{jkl}(p, \eta) \sqrt{p^2 + \tfrac{1}{4}\left|\frac{\partial}{\partial p}\varphi_{jk}(p, \eta)\right|^2}\; e^{-i(\lambda, \varphi_{jk}(p, \eta))}. \tag{4}$$

These formulas and the differentiability properties of the functions $v_{jkl}(p, \eta)$ and $\varphi_j(p, \eta)$ imply the continuous differentiability of the functions $K_{\lambda kl}(p, \eta)$. Putting $p=0$ in formula (4) and using notation (2) we obtain

$$K_{\lambda kl}(0, \eta) = v_{kl}(0, \eta, \eta)\cdot\tfrac{1}{2}\sum_{j=1}^{2}\left|\frac{\partial}{\partial p}\varphi_{jk}(p, \eta)\right|_{p=0}, \tag{5}$$

the last factor in formula (5), by Lemma 1, being different from zero for all $\eta\in[0, H]$.

Applying the operator L defined in Sec. 4 by (4.14) to each equation in (4) we can convert system (4) to the form

$$(u_\lambda(t), K_{\lambda kl}(0, t)) + \int_0^t (u_\lambda(y), R_{\lambda kl}(y, t))\,dy = \mathrm{L}v_{\lambda kl},\qquad k=1,2,\dots,r;\quad l=1,2,\dots,m_k, \tag{6}$$

where

$$R_{\lambda kl}(y, t) = \frac{1}{\pi}\frac{\partial}{\partial t}\int_y^t \frac{K_{\lambda kl}(\sqrt{\eta-y}, \eta)}{\sqrt{(t-\eta)(\eta-y)}}\,d\eta,\qquad k=1,2,\dots,r;\quad l=1,2,\dots,m_k. \tag{7}$$

Consider now matrix $K_\lambda(\eta)$ whose rows are the vectors $K_{\lambda kl}(0, \eta)$, $k=1,2,\dots,r$; $l=1,2,\dots,m_k$. Keeping in mind (5), one can say that matrix $K_\lambda(\eta)$ is formed from the vectors $v_{kl}(0, \eta, \eta)$ $k=1,2,\dots,r$; $l=1,2,\dots,m_k$, with each vector $v_{kl}(0, \eta, \eta)$ multiplied by the non-zero expression $\frac{1}{2}\sum_{j=1}^{2}|(\partial/\partial p)\,\varphi_{jk}(p, \eta)|_{p=0}$.

Since by the hypotheses of the theorem, the vectors $v_{kl}(0, \eta, \eta)$, $k=1,2,\dots,r$; $l=1,2,\dots,m_k$ are independent for all $\eta\in[0, H]$, the determinant of matrix $K_\lambda(\eta)$ is different from zero and it has an inverse $K_\lambda^{-1}(\eta)$. Let $R_\lambda(y, t)$, $V_\lambda(t)$ denote matrices whose rows are made up of the vectors $R_{\lambda kl}(y, t)$ and $Lv_{\lambda kl}$, $k=1, 2, \dots, r;\ l=1, 2, \dots, m_k$, respectively. We are led to a system of Volterra equations of the second kind

$$u_\lambda(t)+\int_0^t K_\lambda^{-1}(t)\, R_\lambda(y, t)\, u_\lambda(y)\, dy=K_\lambda^{-1}(t)\, V_\lambda(t). \tag{8}$$

The kernel $K_\lambda^{-1}(t)\, R_\lambda(y, t)$ of this equation may have a singularity along the diagonal of the same form as the kernel of (4.15). This singularity is easily eliminated by one iteration of the kernel. Thus, the matrix equation (8) has, according to the theory of Volterra equations, a unique solution in the class of continuous functions. From this, we can by repeating the arguments of Sec. 4 convince ourselves of the validity of the formulated theorem.

II. We shall consider, as an application of Theorem 1.7, the problem of determination of an n-dimensional vector-function in the region $\mathscr{D}\{0\leqslant y\leqslant H\}$ of x, y space, $x=(x_1, \dots, x_n)$, $n\geqslant 2$.

Consider n families of curves $L_k(\xi, \eta)$, $k=1, 2, \dots, n$, satisfying conditions 1°–5° of Sec. 4 and take $v_k=dx/ds$ as the weight function for each curve, i.e., consider the system of equations

$$\int_{L_k(\xi, \eta)} (u, dx)=v_k(\xi, \eta) \quad (k=1, 2, \dots, n). \tag{9}$$

The vector dx/ds is the projection of the unit tangent vector to the curve $L_k(\xi, \eta)$ onto the hyperplane $y=0$. It is evident that the differentiability properties of the vector dx/ds are consistent with those required in the theorem formulated above. We shall require the families of curves $L_k(\xi, \eta)$, $k=1, 2, \dots, n$, to be independent in the sense that the unit tangent vectors to the curves $L_k(\xi, \eta)$ constructed at their vertices are independent for any $\eta\in[0, H]$. In this case the conditions of the theorem are fulfilled and the system of equations (9) has a unique solution in the class of continuous functions having compact support in the region $\mathscr{D}$.

6. Determination of a Function in a Circle from its Integrals over a Family of Curves Invariant to Rotation about Center of the Circle

Some inverse problems for differential equations lead (see the chapters to follow) to an integral-geometric problem for families of curves in many respects similar to the one in Sec. 4 with the only difference being that these families are invariant to rotation about some fixed point instead of invariant to parallel translation. To consider such a problem is the purpose of the present section. The method of investigation is similar to that of Sec. 4 with some natural variations. Therefore we shall not go into details but shall concentrate upon those arguments and formulas where changes should be made.

Consider in a unit circle a continuous function $u(r, \varphi)$ and its integrals over a two-parameter family of curves lying inside the circle and invariant to rotation about the center of the circle. It is required to find the function $u(r, \varphi)$ from its integrals. We immediately extend slightly the formulation of the problem by considering it in an arbitrary annulus $\mathscr{D}\{0<r_0\leqslant r\leqslant 1\}$.

Just as in Sec. 4 we shall formulate the geometric limitations on the family of curves under which we are to investigate the integral-geometric problem and we shall ascertain what analytic properties the family of curves has in this case.

Denote by $\boldsymbol{r}$ the radius-vector of a variable point r, φ of a curve of the family and by $\boldsymbol{r}^0$ its unit vector. Let κ be the curvature of the curve at the point r, φ and $\boldsymbol{n}$ the unit normal vector to the curve at this point.

Consider a family of curves satisfying the conditions:

1°. The family of curves is invariant to rotation about the center of a circle.

2°. Each curve of the family is twice continuously differentiable with respect to arclength.

3°. For each point $(\varrho, \alpha)\in\mathscr{D}$ (ϱ, α are polar coordinates of a point) there exists a curve $L(\varrho, \alpha)$, belonging to the family, such that its endpoints lie on the circle $r=1$ and the point (ϱ, α) is its vertex (i.e., the point on the curve which is closest to center of the circle.)

4°. For the family of curves $L(\varrho, \alpha)$ there exist positive numbers $\delta, \varepsilon_1, \varepsilon_2, \kappa_0$ $(\varepsilon_1>\varepsilon_2, \varepsilon_2<1)$ such that the portion of any curve $L(\varrho, \alpha)$ enclosed in the annulus $\mathscr{G}_\varrho=\mathscr{D}\cap\mathscr{D}_\varrho$, $\mathscr{D}_\varrho=\{r-\varrho\leqslant\delta\}$ for any $\varrho\{0<r_0\leqslant \varrho\leqslant 1\}$, represents a continuous curve for which the inequalities

$$0<-\frac{1}{\varrho}+\varepsilon_1\leqslant\kappa\leqslant\kappa_0,\ |(\boldsymbol{r}^0, \boldsymbol{n})|\geqslant\left(\frac{1}{\varrho}-\varepsilon_2\right)r \tag{1}$$

are fulfilled.

5°. The family of curves $L(\varrho, \alpha)$ smoothly fills out the region $\mathscr{D}$ in the sense that the curvature κ of each curve $L(\varrho, \alpha)$ and $(\boldsymbol{r}^0, \boldsymbol{n})$ are continuously differentiable functions of the parameter ϱ.

Analytically conditions 1°–3° mean that the polar equation of the curve $L(\varrho, \alpha)$ can be represented by

$$\begin{aligned}\varphi&=\alpha+\psi(s, \varrho),\\ r&=\varrho+f(s, \varrho),\end{aligned} \tag{2}$$

where $\psi(s, \varrho)$ and $f(s, \varrho)$ are twice continuously differentiable with respect to arclength s (as before, the arclength is measured from the vertex of the curve). Moreover, condition 3° leads to the following restrictions on these functions: $\psi(0, \varrho)=f(0, \varrho)=0$, $f_s(0, \varrho)=0$, $f(s, \varrho)\geqslant 0$. Taking into account that the parameter s is arclength and, hence, that

$$\left|\frac{d\boldsymbol{r}}{ds}\right|^2=\left(\frac{\partial r}{\partial s}\right)^2+r^2\left(\frac{\partial \varphi}{\partial s}\right)^2=1, \tag{3}$$

we find $\psi_s(0, \varrho)=1/\varrho$.

We consider now what restrictions are imposed on the functions ψ, f by condition 4°. Taking into account that the normal vector $\boldsymbol{n}$ is orthogonal to the unit tangent vector $\boldsymbol{\sigma}=d\boldsymbol{r}/ds$, we find

$$|(\boldsymbol{r}^0, \boldsymbol{n})|=r\left|\frac{\partial \varphi}{\partial s}\right|\geqslant r\left(\frac{1}{\varrho}-\varepsilon_2\right). \tag{4}$$

Let us choose now the positive direction of s to be the positive direction of measuring angle φ. Then the above inequality can be written as

$$\frac{\partial \varphi}{\partial s}\geqslant\frac{1}{\varrho}-\varepsilon_2. \tag{4'}$$

In this case the expression for the curvature of the curve has the form

$$\kappa^2=\left|\frac{d^2\boldsymbol{r}}{ds^2}\right|^2=\left[\frac{\partial^2 r}{\partial s^2}-r\left(\frac{\partial \varphi}{\partial s}\right)^2\right]^2+\left[r\frac{\partial^2 \varphi}{\partial s^2}+2\frac{\partial r}{\partial s}\frac{\partial \varphi}{\partial s}\right]^2. \tag{5}$$

We transform the above formula taking into account orthogonality of the vectors $d\boldsymbol{r}/ds$ and $d^2\boldsymbol{r}/ds^2$. The fact that they are orthogonal yields

$$\left(\frac{d\boldsymbol{r}}{ds}, \frac{d^2\boldsymbol{r}}{ds^2}\right)=\frac{\partial r}{\partial s}\left[\frac{\partial^2 r}{\partial s^2}-r\left(\frac{\partial \varphi}{\partial s}\right)^2\right]+r\frac{\partial \varphi}{\partial s}\left[r\frac{\partial^2 \varphi}{\partial s^2}+2\frac{\partial r}{\partial s}\frac{\partial \varphi}{\partial s}\right]=0. \tag{6}$$

Noting that the bracketed expressions in (5) and (6) are identical and using formula (3), we reduce formula (5) to

$$\kappa=\left[\frac{\partial^2 r}{\partial s^2}-r\left(\frac{\partial \varphi}{\partial s}\right)^2\right]\cdot\left[r\frac{\partial \varphi}{\partial s}\right]^{-1}. \tag{7}$$

The first inequality in (1) in this case is equivalent to the inequality

$$r\frac{\partial \varphi}{\partial s}\left[-\frac{1}{\varrho}+\varepsilon_1+\frac{\partial \varphi}{\partial s}\right]\leqslant\frac{\partial^2 r}{\partial s^2}\leqslant\left[\kappa_0+\frac{\partial \varphi}{\partial s}\right]r\frac{\partial \varphi}{\partial s}. \tag{8}$$

We use now inequality (4′) and the inequality $r(\partial\varphi/\partial s)\leqslant 1$ which follows from (3). Then the expressions serving as bounds for the second derivative of r with respect to s, can be estimated as follows:

$$r\frac{\partial\varphi}{\partial s}\left[-\frac{1}{\varrho}+\varepsilon_1+\frac{\partial\varphi}{\partial s}\right]\geqslant(\varepsilon_1-\varepsilon_2)\ r\left(\frac{1}{\varrho}-\varepsilon_2\right)\geqslant(\varepsilon_1-\varepsilon_2)\,(1-\varepsilon_2)=\varepsilon>0,$$

$$\left[\kappa_0+\frac{\partial\varphi}{\partial s}\right]r\frac{\partial\varphi}{\partial s}\leqslant\left(\kappa_0+\frac{1}{r}\right)\leqslant\kappa_0+\frac{1}{r_0}=M<\infty.$$

Thus condition 4° implies that in the region $\mathscr{G}_\varrho$ the function $f(s,\varrho)$ is convex in the variable s, namely

$$0<\varepsilon\leqslant f_{ss}(s,\varrho)\leqslant M<\infty. \tag{9}$$

From condition 5° and formulas (4), (7), (3) and (5) easily follows the continuity of the derivatives

$$\frac{\partial^{k+1}}{\partial\eta\,\partial s^k}\psi(s,\varrho);\qquad\frac{\partial^{k+1}}{\partial\eta\,\partial s^k}f(s,\varrho)\quad(k=1,2)$$

in the region $\mathscr{G}=\cup_{r_0\leqslant\varrho\leqslant 1}\mathscr{G}_\varrho$.

On the basis of the analytical properties of the functions f and ψ and inequality (9) the following lemma is proved analogously to Sec. 4.

Lemma 2. *Let a family of curves satisfy conditions* 1°–5° *in the annulus* $\mathscr{D}\{0<r_0\leqslant r\leqslant 1\}$. *Then in the region* $\mathscr{G}_\varrho$ *the equation of the curve* $L(\varrho,\alpha)$ *can be expressed in the form of two single-valued branches*

$$\varphi=\alpha+\Phi_j(\sqrt{r-\varrho},\varrho),\qquad(j=1,2), \tag{10}$$

the functions $\Phi_j(p,\varrho)$, $j=1,2$, *being continuous in* $\mathscr{G}$ *along with the partial derivatives* $\frac{\partial^{k+l}}{\partial p^k\,\partial\varrho^l}\Phi_j(p,\varrho)$, $k=1,2;\ l=0,1;\ k+l\leqslant 2;\ j=1,2$, *and in addition,*

$$\Phi_j(0,\varrho)=0,\ \left|\varrho\frac{\partial}{\partial p}\Phi_j(p,\varrho)\right|_{p=0}=\sqrt{\frac{2}{f_{ss}(0,\varrho)}}=\sqrt{\frac{2}{M}}>0,\qquad(j=1,2). \tag{11}$$

The theorem analogous to 1.5 is in this case as follows.

Theorem 1.8. *Let* $u(r,\varphi)$ *be a function continuous in the region* $\mathscr{D}\{0<r_0\leqslant r\leqslant 1\}$ *and let the family of curves* $L(\varrho,\alpha)$ *satisfy conditions* 1°–5°. *Then the function* $u(r,\varphi)$ *is uniquely determined by its integrals*

$$\int_{L(\varrho,\alpha)}u(r,\varphi)\,ds=v(\varrho,\alpha),\qquad(\varrho,\alpha)\in\mathscr{D}. \tag{12}$$

The theorem is proved in the same way as Theorem 1.5 with the only difference being that the role of the Fourier transform with respect to the variable α is played by its discrete analog, i.e., by Fourier series. The equation

$$v_n(\varrho)=\int_\varrho^1 u_n(r)\, K_n(\sqrt{r-\varrho},\varrho)\,(r-\varrho)^{-1/2}\,dr \quad (n=0,\pm 1,\pm 2,\ldots), \tag{13}$$

is obtained for each coefficient $u_n(r)$ of the Fourier series of the function $u(r,\varphi)$. Eq. (13) is fully analogous to (4.12). Here the analogue of (4.11) is the formula

$$K_n(p,\varrho)=\sum_{j=1}^{2}\sqrt{p^2+\tfrac{1}{4}(\varrho+p^2)^2\left|\frac{\partial}{\partial p}\Phi_j(p,\varrho)\right|^2}\; e^{-in\Phi_j(p,\varrho)}, \tag{14}$$
$$(n=0,\pm 1,\pm 2,\ldots),$$

and also

$$K_n(0,\varrho)=\tfrac{1}{2}\sum_{j=1}^{2}\varrho\left|\frac{\partial}{\partial p}\Phi_j(p,\varrho)\right|_{p=0}=\sqrt{\frac{2}{\psi_{ss}(0,\varrho)}}\geqslant\sqrt{\frac{2}{M}}>0. \tag{15}$$

The differentiability properties of the functions $K_n(p,\varrho)$ and $K_\lambda(p,\eta)$ are fully identical. Therefore in order to obtain a Volterra equation of second kind from (13), it is sufficient to apply to (13) the operator L given by

$$\mathrm{L}v_n=\frac{1}{\pi}\frac{d}{dt}\int_t^1\frac{v_n(\varrho)\,d\varrho}{\sqrt{\varrho-t}}. \tag{16}$$

As a result of its application, we obtain the equation

$$u_n(t)+\int_t^1 u_n(r)\,R_n(r,t)\,dr=-\frac{1}{K_n(0,t)}\mathrm{L}v_n \qquad (n=0,\pm 1,\pm 2,\ldots), \tag{17}$$

where

$$R_n(r,t)=-\frac{1}{\pi\,K_n(0,t)}\frac{\partial}{\partial t}\int_t^r\frac{K_n(\sqrt{r-\varrho},\varrho)}{\sqrt{(r-\varrho)(\varrho-t)}}\,d\varrho. \tag{18}$$

In terms of its properties, (17) is fully analogous to (4.15). Therefore repeating the reasoning of Theorem 1.5, we can be convinced of the validity of the above theorem.

Certainly, it is possible, just as in Sec. 4, to pose a more general problem for the function $u(r, \varphi)$ by specifying the integrals over the family of curves $L(\varrho, \alpha)$ with a weight function $q(r, \varrho, \varphi-\alpha)$ that is continuously differentiable in all arguments and different from zero at the vertex of each curve $L(\varrho, \alpha)$, i.e., such that $q(\varrho, \varrho, 0) \neq 0 \, (0 < r_0 \leqslant r \leqslant 1)$. A uniqueness theorem then holds.

7. Integral-Geometric Problem for Surfaces Invariant to Displacement

The investigations of Secs. 4–5 were concerned with problems of finding one or more functions from their integrals over a family of curves invariant to parallel translation along a hyperplane. It is not difficult to extend the results obtained in Secs. 4–5 to the case where instead of the integrals over the curves, the integrals are specified over a family of surfaces invariant to all parallel translations along a fixed hyperplane. In what follows, we shall consider these questions.

I. Consider in $(n+1)$-dimensional x, y space, $(n \geqslant 2)$, $x=(x_1, \ldots, x_n)$, the region $\mathscr{D}\{0 \leqslant y \leqslant H\}$ and in it a family of hypersurfaces (i.e. n-dimensional surfaces) which depends on an $(n+1)$st parameter. Let the family satisfy the following conditions:

1°. The family of surfaces is invariant to all parallel translations along the hyperplane $y=0$.

2°. For each point $(\xi, \eta) \in \mathscr{D}$, $\xi=(\xi_1, \ldots, \xi_n)$, there exists a hypersurface $S(\xi, \eta)$ belonging to the family such that it has the form of a cap whose vertex is (ξ, η) and base lies in the hyperplane $y=0$.

3°. The equation of the surface $S(\xi, \eta)$ can be represented in a form with the variable y solved for, i.e. taking into account conditions 1°–2°, in the form

$$y=\eta-f(x-\xi, \eta), \tag{1}$$

where $f(x-\xi, \eta)$ is continuous in the region $\mathscr{D}$ along with its partial derivatives in the variable x up to order $\nu=[(n+5)/2]$ inclusive.

4°. For the family of surfaces $S(\xi, \eta)$ there exist positive numbers δ, m, M such that the portion of any $S(\xi, \eta)$ enclosed in the region $\mathscr{G}_\eta=\mathscr{D} \cap \mathscr{D}_\eta$, $\mathscr{D}_\eta=\{\eta-y \leqslant \delta\}$ for all $\eta \in [0, H]$ is a simply connected strictly convex surface:

$$m|x-\xi|^2 \leqslant \sum_{i,j=1}^{n} \frac{\partial^2}{\partial x_i \, \partial x_j} f(x-\xi, \eta)\,(x_i-\xi_i)\,(x_j-\xi_j) \leqslant M|x-\xi|^2. \tag{2}$$

5°. The family of surfaces smoothly fills out the region $\mathscr{D}$ in the sense that the partial derivatives

$$\frac{\partial^{k+l+m}}{\partial x_i^k \, \partial x'_j \, \partial \eta^m} f(x-\xi, \eta),$$

$$k, l=0, 1, \ldots, \nu; \; m=0, 1, 2, \ldots, \nu-1; \; k+l \leqslant \nu; \; k+l+m \leqslant \nu+1,$$

are continuous in $\mathcal{D}$ in all arguments.

In our further investigation it will be convenient to represent the equation of the surface $S(\xi, \eta)$ in cyclindrical coordinates, which we introduce as follows: for one of the variables we take y and at the point ξ in $x_1, \ldots, x_n$-space we introduce spherical coordinates by the formulas

$$\begin{aligned} x_1 &= \xi_1 + r \cos \varphi_1, \\ x_2 &= \xi_2 + r \sin \varphi_1 \cos \varphi_2, \\ x_3 &= \xi_3 + r \sin \varphi_1 \sin \varphi_2 \cos \varphi_3, \\ &\cdots\cdots\cdots\cdots \\ x_{n-1} &= \xi_{n-1} + r \sin \varphi_1 \sin \varphi_2 \cdot \ldots \cdot \sin \varphi_{n-2} \cos \varphi_{n-1}, \\ x_n &= \xi_n + r \sin \varphi_1 \sin \varphi_2 \cdot \ldots \cdot \sin \varphi_{n-2} \sin \varphi_{n-1}. \end{aligned} \tag{3}$$

The domain of variation of the angles $\varphi_1, \ldots, \varphi_{n-1}$ is the parallelepided $\mathcal{Q}\{0 \leqslant \varphi_i \leqslant \pi, \, i=1, 2, \ldots, n-2; \, 0 \leqslant \varphi_{n-1} \leqslant 2\pi\}$.

The following lemma is valid.

Lemma 3. *In cylindrical coordinates $y, r, \varphi, \varphi=(\varphi_1, \ldots, \varphi_{n-1})$, the equation of the surface* (1) *can be represented in the region in the form*

$$r = F(\sqrt{\eta - y}, \eta, \varphi), \tag{4}$$

$F(p, \eta, \varphi)$ being continuous in the region $\mathcal{G}(0 \leqslant p \leqslant \delta, 0 \leqslant \eta \leqslant H, \varphi \in \mathcal{Q}\}$ along with its partial derivatives

$$\frac{\partial^{k+l+m}}{\partial p^k \, \partial \eta^l \, \partial \varphi_i^m} F(p, \eta, \varphi), \qquad k, l=0, 1, 2, \ldots, \nu; \; m=0, 1, 2, \ldots, \nu-1;$$

$$k+l \leqslant \nu, \; k+l+m \leqslant \nu,$$

and such that

$$F(0, \eta, \varphi)=0; \; \left| \frac{\partial}{\partial p} F(p, \eta, \varphi) \right|_{p=0} \geqslant \sqrt{\frac{2}{M}} > 0. \tag{5}$$

The proof of the lemma is only slightly different from that of Lemma 1. It will be noted, first of all, that by condition 2°, $f(0, \eta)=0$, $f_{x_i}(0, \eta)=0$, $i=1, 2, \ldots, n$. We denote now by $g(r, \eta, \varphi)$ the function obtained as a result of substituting in the function $f(x-\xi, \eta)$ in place of the variable $x-\xi$, its expression in terms of r, φ according to formulas (3) and we compute its partial derivatives with respect to the variable r. By the chain rule, we find

$$g_r(r, \eta, \varphi) = \sum_{i=1}^{n} f_{x_i} \frac{\partial x_i}{\partial r}, \tag{6}$$

$$g_{rr}(r, \eta, \varphi) = \sum_{i,j=1}^{n} f_{x_i x_j} \frac{\partial x_i}{\partial r} \frac{\partial x_j}{\partial r} = \sum_{i,j=1}^{n} f_{x_i x_j} \frac{(x_i - \xi_i)(x_j - \xi_j)}{|x - \xi|^2}.$$

From the first formula we obtain $g_r(0, \eta, \varphi) = 0$, while from $f(0, \eta) = 0$ there follows $g(0, \eta, \varphi) = 0$. Denote by $\mathscr{T}_{\xi,\eta}$ the region of $x_1, \ldots, x_n$ space which is the domain of the function $f(x-\xi, \eta)$ assuming that its set of values represents a segment $[0, \delta]$. Then the second formula in (6) in conjunction with condition 4° allows us to state that in the region $\tilde{\mathscr{T}}_{\eta}$, which is the image of the region $\mathscr{T}_{,\eta}$ in the variables r, φ, the inequality

$$0 < m \leqslant g_{rr}(r, \eta, \varphi) \leqslant M \tag{7}$$

holds for all $\eta \in [0, H]$. But in that case the equation

$$y = \eta - g(r, \eta, \varphi) \tag{8}$$

in the region $\tilde{\mathscr{T}}_{\eta}$ can be solved with respect to the variable r and it has form (4) as follows from the arguments of Lemma 1. The continuity of the derivatives of the required $F(p, \eta, \varphi)$ is easily established by using as in Lemma 1, (8) and the differentiability properties of function $f(x-\xi, \eta)$ mentioned in 3°, 5°. The inequality in formula (5) is proved as follows. Rewrite equality (8) in the form

$$p = \sqrt{g(r, \eta, \varphi)}, \quad (p = \sqrt{\eta - y}), \tag{9}$$

and differentiate it with respect to p assuming $r = F(p, \eta, \varphi)$. We then obtain the equality

$$F_p(p, \eta, \varphi) = \left. \frac{2\sqrt{g(r, \eta, \varphi)}}{g_r(r, \eta, \varphi)} \right|_{r = F(p, \eta, \varphi)} \tag{10}$$

Removing the indeterminacy on the right hand side arising as $p \to 0$, we find

$$F_p(0, \eta, \varphi) = \sqrt{\frac{2}{g_{rr}(0, \eta, \varphi)}} \geqslant \sqrt{\frac{2}{M}},$$

as required. The property $F(0, \eta, \varphi) = 0$ follows from the equality $g(0, \eta, \varphi) = 0$ and formula (9) from which we find that the value $r = 0$ corresponds to the value $p = 0$.

Hence, Lemma 3 is proved.

II. We now proceed to state the results for the integral-geometric problems in the case of surfaces. In many respects they are analogous to the results of Secs. 4–5.

Theorem 1.9 *Let $u(x, y)$ be a continuous function with compact support in the region $\mathscr{D}$ and let the family of surfaces $S(\xi, \eta)$, $(\xi, \eta)\in\mathscr{D}$, satisfy conditions* 1°–5°. *Then the function $u(x, y)$ is uniquely determined by its integrals*

$$\iint_{S(\xi,\eta)} u(x, y)\, d\sigma = v(\xi, \eta), \qquad (\xi, \eta)\in\mathscr{D}, \tag{11}$$

in which $d\sigma$ is the element of area of the hypersurface $S(\xi, \eta)$ and the integration is carried out over the whole hypersurface $S(\xi, \eta)$.

Proof. Rewrite equality (11), using the invariance of the family of surfaces $S(\xi, \eta)$ to displacement along the hyperplane $y=0$, in the form

$$\iint_{S(0,\eta)} u(x+\xi, y)\, d\sigma = v(\xi, \eta). \tag{11'}$$

Taking Fourier transforms of both sides in the variable ξ, we have

$$\iint_{S(0,\eta)} u_\lambda(y)\, e^{-i(\lambda, x)}\, d\sigma = v_\lambda(\eta). \tag{12}$$

Here $u_\lambda(y)$ and $v_\lambda(\eta)$ are, as usual, the Fourier transforms of $u(x, y)$ and $v(\xi, \eta)$. Change now the surface integral in formula (12) selecting y to be one of the variables of integration. Denote by $\Sigma(y_0, \eta)$ the surface of the cross-section of the surface $S(0, \eta)$ by the hyperplane $y=y_0$. Let $d\Sigma$ be element of area of the surface $\Sigma(y, \eta)$ and let γ be the angle between the normal to the surface $S(0, \eta)$ at the point (x, y) and the axis y. Then the element of surface area $d\sigma$ can be written as

$$d\sigma = \frac{dy\, d\Sigma}{|\sin\gamma|}. \tag{13}$$

The integral equation (12) now takes the form of a Volterra equation of the first kind

$$\int_0^\eta u_\lambda(y)\, K_\lambda(y, \eta)\, dy = v_\lambda(\eta) \qquad (0 \leqslant \eta \leqslant H) \tag{14}$$

with kernel given by the formula

$$K_\lambda(y, \eta) = \iint\limits_{\Sigma(y, \eta)} e^{-i(\lambda, x)} \frac{d\Sigma}{|\sin \gamma|}. \tag{15}$$

Let us study the properties of the kernel of the equation. As follows from the arguments of Theorem 1.5, it is sufficient to study its properties in a δ-neighbourhood of the diagonal, i.e. in the region $0 \leqslant \eta - y \leqslant \delta$. But in this case we can make use of Lemma 3 and represent the equation of the portion of the surface falling in the region $0 \leqslant \eta - y \leqslant \delta$ in cylindrical coordinates

$$r = F(\sqrt{\eta - y}, \eta, \varphi). \tag{4}$$

Keeping the variables η, y fixed, we obtain the polar equation of the surface $\Sigma(y, \eta)$. Let us compute the sine of the angle between the normal to the surface $S(0, \eta)$ and the y-axis and the element of area of the surface $d\Sigma$ in formula (15). To compute $\sin\gamma$, we note that for fixed φ, Eq. (4) is the equation of a curve lying on the surface $S(0, \eta)$, and moreover, this curve lies in the two-dimensional plane containing the y axis and the straight line in the hyperplane $y=0$ passing through the origin and determined by the angular coordinate φ. Owing to this, it is sufficient to find the sine of the angle between the normal to the curve and the y axis or, equivalently in this case, the cosine of the angle between the tangent to the curve and the y axis. The latter is easily calculated by use of the parametric equations of the curve obtained from formulas (3) by setting $\xi_i = 0$, $i = 1, 2, \ldots, n$, and r equal to the right hand side of formula (4). Carrying out the necessary calculations we obtain

$$\sin\gamma = \frac{1}{\sqrt{1 + |x_y|^2}} = \frac{1}{\sqrt{1 + r_y^2}}. \tag{16}$$

To calculate the element of surface area $d\Sigma$, we use the parametric equations of the surface $\Sigma(y, \eta)$ which we obtain by substituting the expression for r from formula (4) into (3) and putting $\xi_1 = \xi_2 = \cdots = \xi_n = 0$. The element of surface area is obtained by forming the matrix

$$A = \left\| \frac{\partial x_i}{\partial \varphi_j} \right\| \qquad \begin{array}{l} i = 1, 2, \ldots, n; \\ j = 1, 2, \ldots, n-1, \end{array}$$

and computing the determinant of matrix AA^* where A^* is the transpose of A. The element of surface area of $\Sigma(y, \eta)$ is thus given by the formula

$$d\Sigma = \sqrt{|A\, A^*|}\, d\varphi_1 \ldots d\varphi_{n-1}. \tag{17}$$

In our case the formula

$$\sqrt{|A\,A^*|}=r^{n-2}\sqrt{r^2+\sum_{i=1}^{n-1}\frac{r_{\varphi_i}^2}{q_i^2}\prod_{i=1}^{n-2}\sin^{(n-i-1)}\varphi_i} \tag{18}$$

holds, where

$$q_i=\begin{cases}1 & i=1,\\ \prod_{k=1}^{i-1}\sin\varphi_k, & i=2,3\,\ldots,n-1.\end{cases} \tag{19}$$

Now, relying on formulas (16) and (18), we write the kernel of (14) in the form of multiple integral

$$K_\lambda(y,\eta)\,|_{y=\eta-p^2}\equiv K_{1\lambda}(p,\eta)=\int_0^\pi d\varphi_1\ldots\int_0^\pi d\varphi_{n-2}\int_0^{2\pi}d\varphi_{n-1}\,e^{-i(\lambda,x)}$$
$$\cdot\frac{\sqrt{p^2+\frac{1}{4}F_p^2(p,\eta,\varphi)}}{p}\,F^{n-2}(p,\eta,\varphi)$$
$$\cdot\sqrt{F^2(p,\eta,\varphi)+\sum_{i=1}^{n-1}\frac{F_{\varphi_i}^2(p,\eta,\varphi)}{q_i^2}}\cdot\prod_{i=1}^{n-2}\sin^{n-i-1}\varphi_i. \tag{20}$$

Of course, in this formula, $x_1,\ldots,x_n$ should be replaced by their expressions in terms of p,η,φ in (3) in which we set $\xi_1=\xi_2=\cdots=\xi_n=0$, $r=F(p,\eta,\varphi)$. Making use of formula (20) we shall show that the function $K_{1\lambda}(p,\eta)$ can be represented in the form

$$K_{1\lambda}(p,\eta)=p^{n-2}K_{2\lambda}(p,\eta), \tag{21}$$

where the function $K_{2\lambda}(p,\eta)$ is continuous in the region $\{0\leqslant p\leqslant\delta, 0\leqslant\eta\leqslant H\}$ along with its partial derivatives $(\partial^{k+m}/\partial p^k\,\partial\eta^m)\,K_{2\lambda}(p,\eta)$, $m,k=0,1,2,\ldots,\nu-2$, $k+m\leqslant\nu-2$.

Indeed, using (21) to express $K_{2\lambda}(p,\eta)$ in terms of $K_{1\lambda}(p,\eta)$, we find

$$K_{2\lambda}(p,\eta)=\int_0^\pi d\varphi_1\ldots\int_0^\pi d\varphi_{n-2}\int_0^{2\pi}d\varphi_{n-1}\,e^{-i(\lambda,x)}$$
$$\cdot\sqrt{p^2+\tfrac{1}{4}F_p^2(p,\eta,\varphi)}\left[\frac{1}{p}F(p,\eta,\varphi)\right]^{n-2} \tag{22}$$
$$\cdot\sqrt{\left[\frac{1}{p}F(p,\eta,\varphi)\right]^2+\sum_{i=1}^{n-1}\left[q_i^{-1}\frac{1}{p}F_{\varphi_i}\right]^2}\prod_{i=1}^{n-2}\sin^{n-i-1}\varphi_i$$

It will be remembered that by Lemma 3, $F(0, \eta, \varphi)=0$ and $F_p(p, \eta, \varphi)|_{p=0} \geqslant \sqrt{(2/m)}>0$. From these properties of $F(p, \eta, \varphi)$ and the smoothness of its derivatives it follows that the expressions $(1/p)\, F(p, \eta, \varphi)$ and $(1/p)\, F_{\varphi_i}(p, \eta, \varphi)$ are continuous along with the partial derivatives, generally, up to an order one less than the order of smoothness of $F(p, \eta, \varphi)$, $F_{\varphi_i}(p, \eta, \varphi)$ themselves. Taking account of the above as well as the degree of smoothness of $F(p, \eta, \varphi)$, we can be convinced of the smoothness of the function $K_{2\lambda}(p, \eta)$ mentioned above. In addition, it will be noted that from (22) follows the inequality

$$K_{2\lambda}(0, \eta) \geqslant \tfrac{1}{2} \int_0^{\pi} d\varphi_1 \ldots \int_0^{\pi} d\varphi_{n-2} \int_0^{2\pi} d\varphi_{n-1} [F_p(0, \eta, \varphi)]^{n-1} \prod_{i=1}^{n-1} \sin^{n-i-1} \varphi_i$$

$$\geqslant \left(\frac{2\pi}{M}\right)^{n/2} \frac{1}{\Gamma(n/2)} > 0. \tag{23}$$

Thus (14) can be written as

$$\int_0^{\eta} u_\lambda(y)\, (\eta - y)^{(n-2)/2}\, K_{2\lambda}(\sqrt{\eta - y}, \eta)\, dy = v_\lambda(\eta). \tag{24}$$

Strictly speaking, what we have written is not quite true because we have shown that the kernel of (14) is representable in form (21) only in a δ-neighbourhood of the diagonal $(0 \leqslant \eta - y \leqslant \delta, 0 \leqslant y \leqslant H)$, i.e., formula (24) is true only for $0 \leqslant \eta \leqslant \delta$. However, in Sec. 4 we gave the reasoning by which it follows that in investigating the equations of type (14) we may proceed by δ-steps in η performing the evaluation of the integrals from the solution already obtained, so that we shall need only a δ-neighbourhood of the diagonal. Therefore we do not stipulate specifically the bounds of variation of the variable η. It is not difficult, having equation (24), to obtain a Volterra equation of the second kind.

In the case of even n it is sufficient to differentiate equality (24) $n/2 = \nu - 2$ times in order to arrive at the equation

$$(\nu - 3)!\, K_{2\lambda}(0, \eta)\, u_\lambda(\eta) + \int_0^{\eta} u_\lambda(y)\, T_\lambda(y, \eta)\, dy = \frac{d^{\nu-2}}{d\eta^{\nu-2}} v_\lambda(\eta), \tag{25}$$

in which

$$T_\lambda(y, \eta) = \frac{\partial^{\nu-2}}{\partial \eta^{\nu-2}} [(\eta - y)^{\nu-3}\, K_{2\lambda}(\sqrt{\eta - y}, \eta)]. \tag{26}$$

Eq. (25) after division by the expression $(\nu-3)!\,K_{2\lambda}(0,\eta)$ becomes a usual Volterra equation of the second kind with a weak singularity.

In the case of odd n, (24) must be differentiated $(n-1)/2=\nu-3$ times and then the operator L defined by (4.14) applied. Carrying out these operations, we arrive at the equation

$$u_\lambda(t)+\int_0^t u_\lambda(y)\,Q_\lambda(y,t)\,dy=\frac{1}{K_{2\lambda}(0,t)}\,\frac{2^{(n-1)/2}}{1\cdot 3\cdot 5\ldots(n-2)}\,\mathrm{L}\left(\frac{d^{\nu-3}}{d\eta^{\nu-3}}\,v_\lambda(\eta)\right), \tag{27}$$

whose kernel is given by

$$Q_\lambda(y,t)=\frac{1}{\pi K_{2\lambda}(0,t)}\cdot\frac{2^{(n-1)/2}}{1\cdot 3\cdot 5\cdot\ldots\cdot(n-2)}\cdot\frac{\partial}{\partial t}\int_y^t\frac{\frac{\partial^{\nu-3}}{\partial\eta^{\nu-3}}\left[K_{2\lambda}(\sqrt{y-\eta},\eta)\,(\eta-y)^{(2(\nu-3)-1)/2}\right]}{\sqrt{(t-\eta)\,(\eta-y)}}\,d\eta \tag{28}$$

Generally speaking the kernel of (27) has a weak polarity just as in the previous case.

So, for any n we can reduce (14) to a Volterra equation of the second kind. From this follows the validity of Theorem 1.9.

III. Just as in the case of curves one can consider an integral-geometric problem with known weight function $\varrho(x-\xi,y,\eta)$. If, instead of integrals (11) we consider

$$\iint_{S(\xi,\eta)}\varrho(x-\xi,y,\eta)\,u(x,y)\,d\sigma=v(\xi,\eta),\qquad(\xi,\eta)\in\mathscr{D}, \tag{29}$$

then, as before, we arrive at (14) with kernel of the form

$$K_\lambda(y,\eta)=\iint_{\Sigma(y,\eta)}\varrho(x,y,\eta)\,e^{-i(\lambda,x)}\,\frac{d\Sigma}{|\sin\gamma|}. \tag{15'}$$

All calculations of Theorem 1.9 remain valid if we require continuous differentiability of the function $\varrho(x-\xi,y,\eta)$ in all arguments up to order $\nu-2$ and the fulfillment of the condition $\varrho(0,\eta,\eta)\neq 0$, $0\leqslant\eta\leqslant H$.

Thus if these conditions are fulfilled, (29) has a unique solution in the class of the continuous functions having compact support.

Analogously to Sec. 6 one can consider an m-dimensional vector function $u(x,y)$ and the integral-geometric problem for it. Let there be r

families of surfaces $S_k(\xi, \eta)$ $(k=1, 2, \ldots, r)$ and let each of them generate m_k scalar equations for $u(x, y)$:

$$\iint\limits_{S_k(\xi,\eta)} (\varrho_{kl}, u)\, d\sigma = v_{kl}(\xi, \eta) \tag{30}$$

$$\begin{array}{l} k=1, 2, \ldots, r \;\; ; \\ l=1,2, \ldots, m_k\, ; \end{array} \qquad (m_1+m_2+\cdots+m_r=m).$$

A theorem completely analagous to Theorem 1.7 is valid.

Theorem 1.10. *Let the vector function $u(x, y)$ have compact support and be continuous in the region $\mathscr{D}\{0\leqslant y\leqslant H\}$, the families of surfaces $S_k(\xi, \eta)$ satisfy conditions 1°–5° and the weight functions $\varrho_{kl}\equiv\varrho_{kl}(x-\xi, y, \eta)$, $k=1, 2, \ldots, r$; $l=1, 2, \ldots, m_k$, be continuous along with their partial derivatives up to the order $\nu-2$ and such that the system of vectors $\varrho_{kl}(0, \eta, \eta)$ $k=1, 2, \ldots, r$; $l=1, 2, \ldots, m_k$, is linearly independent for any $\eta\in[0, H]$. Then the vector function $u(x, y)$ is uniquely determined by system* (30).

We shall not discuss the proof of this theorem because it is just a paraphrasal of that of Theorem 1.7 to the case of surfaces taking into account the analysis carried out above for one equation of form (30).

It will also be noted that instead of integrals of form (29) one can pose an integral-geometric problem by specifying a linear combination of integrals with different weights over the surfaces $S(\xi, \eta)$ and volumes $V(\xi, \eta)$ bounded by them as follows:

$$\iint\limits_{S(\xi,\eta)} \varrho(x-\xi, y, \eta)\, u(x, y)\, d\sigma + \iiint\limits_{V(\xi,\eta)} \varrho_1(x-\xi, y, \eta)\, u(x, y)\, dv = v(\xi, \eta), \quad (\xi, \eta)\in\mathscr{D}. \tag{31}$$

Here dv is the volume element. Regarding the family of surfaces $S(\xi, \eta)$ and the function $\varrho(x, y, \eta)$, we shall suppose, that they satisfy the above conditions and that $\varrho_1(x, y, \eta)$ is continuous together with its partial derivatives up to the order $\nu-2$. Then the function $u(x, y)$ is uniquely determined by the function $v(\xi, \eta)$. This statement is easily verified by writing out the kernel of (14) corresponding to (31). The order of singularity of (14) near the diagonal is determined in this case by the surface integral; the volume integral introduces higher order singularity as $y\to\eta$.

8. Integral-Geometric Problems for a Family of Curves Generated by a Riemannian Metric

In the following chapters, we shall investigate inverse problems for second order linear hyperbolic equations of the form

$$u_{tt}=\sum_{i,j=1}^{m} a_{ij}(x)\, u_{x_i x_j}+Su+f(x,t). \tag{1}$$

Here S is a first order linear differential operator, $x=(x_1,\dots,x_m)$, and the coefficients $a_{ij}(x)$ satisfy for any vector $\alpha=(\alpha_1,\dots,\alpha_m)$ the inequality

$$m\,|\alpha|^2\leqslant\sum_{i,j=1}^{m} a_{ij}(x)\,\alpha_i\alpha_j\leqslant M\,|\alpha|^2, \qquad (0<m\leqslant M<\infty). \tag{2}$$

The coefficients of (1) will be assumed to possess certain smoothness properties. In particular the $a_{ij}(x)$ are to belong to C^2. Some of the inverse problems for (1) reduce to the problems considered in this chapter. At the same time they have their own special character in that the integrals are given over a family of curves which are geodesics in a Riemannian metric related to the differential equation (1) in a natural manner. It is known that for (1) an important role is played by the curves $\Gamma(x,\xi)$ on which the functional

$$\tau(L)=\int\limits_{L(x,\xi)}\sqrt{\sum_{i,j=1}^{m} b_{ij}(x)\,dx_i\,dx_j}, \tag{3}$$

has an extremum, where $L(x,\xi)$ is any curve joining the fixed points x, ξ, $\xi=(\xi_1,\dots,\xi_m)$, and the b_{ij} are elements of matrix $\bar{B}$ inverse to symmetric matrix $\bar{A}=\|a_{ij}\|$, $i,j=1,2,\dots,m$. In other words the curves $\Gamma(x,\xi)$ are geodesics in the metric in which the element of arclength ds is given by

$$ds=\sqrt{\sum_{i,j=1}^{m} b_{ij}(x)\,dx_i\,dx_j}. \tag{4}$$

In the literature the geodesics $\Gamma(x,\xi)$ are often termed the extremals or rays of metric (4) and we shall use this terminology.

I. The purpose of this section is to find conditions on the metric coefficients of (4) under which the integral-geometric problem over a family of geodesics has a unique solution. We confine ourselves to the case where the metric coefficients are functions of a single coordinate. Let x_m be this coordinate. For convenience it will be denoted by y and the other coordinates will be denoted by $x_1,\dots,x_m$, i.e. we put $m=n+1$. Accordingly the coefficient of $u_{x_i x_j}(i,j=1,2,\dots,n)$ in (1) will again be denoted a_{ij} and new notation $q_i(i=1,2,\dots,n)$ and s will be introduced for the coefficients of $u_{x_i y}(i=1,2,\dots,n)$ and u_{yy}. Then the coefficient matrix of the higher order derivatives will have the following structure:

$$\bar{A}=\begin{pmatrix} A & q^* \\ q & s \end{pmatrix}, \quad \begin{aligned} &A=\|a_{ij}\| \, (i,j=1,2,\dots,n), \\ &q=\|a_{n+1\; j}\| \, (j=1,2,\dots,n), \\ &s=a_{n+1 n+1}. \end{aligned} \tag{5}$$

Here q^* is the column which is the transpose of the row matrix q. For the coefficients b_{ij} we introduce similar notation:

$$\bar{B}=\begin{pmatrix} B & p^* \\ p & r \end{pmatrix}, \quad \begin{aligned} &B=\|b_{ij}\| \; (i,j=1,2,\dots,n), \\ &p=\|b_{n+1\; j}\| \; (j=1,2,\dots,n), \\ &r=b_{n+1\;\, n+1}. \end{aligned} \tag{6}$$

Then the metric to be considered is given in matrix notation by

$$ds=\sqrt{dx\, B(y)\, dx^*+2p(y)\, dx^*\, dy+r(y)\, dy^2}. \tag{7}$$

It is obvious that the family of geodesics generated by metric (7) is invariant to displacement along the hyperplane $y=0$. Therefore we consider the halfspace $y\geqslant 0$ and we try to find what conditions on the coefficients of metric (7) assure regularity of behaviour of the geodesics in this region necessary for the integral-geometric problems. We make use of Euler's equations for the extremals of functional (3) and we write out explicitly the equations for geodesics. For this purpose we denote $dx/dy=x_y$ on the geodesic and

$$\Delta=\sqrt{x_y B x_y^*+2p\, x_y^*+r}\,(\operatorname{sgn} dy). \tag{8}$$

The expression Δ is positive at those points of the geodesic for which the coordinate y increases in moving along the geodesic in the positive direction and is negative if the coordinate y decreases. In this case Euler's equations [71] determining the geodesic lines have the form

$$\operatorname{grad}_x \Delta-\frac{d}{dy}\operatorname{grad}_{x_y}\Delta=0. \tag{9}$$

The subscript on the symbol grad indicates with respect to what variables the gradient of Δ is to be computed. Noting that Δ does not depend explicitly on x, we find from (9) that the expression

$$\operatorname{grad}_{x_y}\Delta=\frac{x_y B+p}{\Delta}=z \tag{10}$$

remains constant along each geodesic of metric (7). The parameter $z=(z_1,\dots,z_n)$ is determined at a point by specifying the direction cosines of the tangent to the geodesic and, hence, it is an arbitrary parameter of the family of geodesics. We now express x_y in terms of the metric coefficients and the parameter z. For this purpose we find x_y in terms of z

and Δ and substitute in formula (8). To compute Δ in this case we obtain the following formula:

$$\Delta=\sqrt{\frac{r-p\,B^{-1}p^*}{1-zB^{-1}z^*}}\operatorname{sgn} dy. \tag{11}$$

Substituting now the expression for Δ into formula (10), we find the differential equation for geodesics to be

$$x_y=-p(y)\,B^{-1}(y)+zB^{-1}(y)\,\Delta. \tag{12}$$

Integrating (12) we obtain an explicit equation for the geodesics crossing the hyperplane $y=0$ at $(x^0, 0)$

$$x=x^0+\int\limits_0^y[-p(y)\,B^{-1}(y)+z\,B^{-1}(y)\,\Delta]\,dy \tag{13}$$

and depending on $2n$ arbitrary constants $z_1,\dots,z_n$, $x_1^0,\dots,x_n^0$.

It will be noted that matrix B^{-1} and the expression $r-pB^{-1}p^*$ occurring in formula (11) are related very simply to the elements of matrix $\bar{A}$; namely,

$$\left.\begin{aligned} B^{-1}&=A-\frac{1}{s}q^*q,\\ r-pB^{-1}p^*&=\frac{1}{s}\end{aligned}\right\}. \tag{14}$$

These formulas are easily obtained if one takes into account that $\bar{B}$ is the inverse of matrix $\bar{A}$ and, hence, the following equations are valid:

$$\left.\begin{aligned} AB+q^*p&=I,\\ qB+sp&=0,\\ qp^*+rs&=I,\\ Ap^*+rq^*&=0\end{aligned}\right\}, \tag{15}$$

in which I is the identity matrix. Note further that (15) implies also $-pB^{-1}=q/s$ and therefore the equation of geodesics (13) can be written as

$$x=x^0+\int\limits_0^y\left[\frac{q(y)}{s(y)}+zB^{-1}(y)\,\Delta\right]dy, \tag{13'}$$

where

$$\Delta = \frac{\operatorname{sgn} dy}{\sqrt{s(y)\,[1 - zB^{-1}(y)\,z^*]}}. \tag{11'}$$

Let us point out some restrictions on the coefficient $s(y)$ and matrix $B^{-1}(y)$ following from condition (2). First of all setting $\alpha_{n+1}=1$, $\alpha_i=0$, $i=1, 2, \ldots, n$, we obtain the restriction on $s(y)$:

$$0 < m \leqslant s(y) \leqslant M < \infty. \tag{16}$$

We show now that the matrix $B^{-1}(y)$ satisfies an inequality analogous to (2) for any real vector $\beta = (\beta_1, \ldots, \beta_n)$, namely

$$m\,\beta\,\beta^* \leqslant \beta B^{-1}(y)\,\beta^* \leqslant M\beta\,\beta^*. \tag{17}$$

Indeed, using (14), we find

$$\beta B^{-1}\beta^* = \beta A\beta^* - \frac{1}{s}(\beta\, q^*)^2. \tag{18}$$

On the other hand, setting $\alpha = (\beta, \alpha_{n+1})$ in inequality (2), we have

$$m[\beta\beta^* + \alpha_{n+1}^2] \leqslant \beta A\,\beta^* + 2(\beta q^*)\,\alpha_{n+1} + s\alpha_{n+1}^2 \leqslant M[\beta\,\beta^* + \alpha_{n+1}^2]. \tag{19}$$

Hence, rewriting the first half of inequality (19) in the form of an inequality for quadratic trinomial in α_{n+1}, namely,

$$(\beta A\beta^* - m\beta\beta^*) + 2(\beta q^*)\,\alpha_{n+1} + (s-m)\,\alpha_{n+1}^2 \geqslant 0,$$

we conclude because α_{n+1} is arbitrary that

$$(\beta q^*)^2 - (s-m)\,(\beta A\beta^* - m\,\beta\,\beta^*) \leqslant 0,$$

or, to put it differently, dividing by s,

$$\beta A\beta^* - \frac{1}{s}(\beta q^*)^2 \geqslant \frac{1}{s}(\beta A\,\beta^* - m\,\beta\beta^*) + m\beta\beta^*. \tag{20}$$

But from inequality (2), we find setting $\alpha = (\beta, 0)$ that

$$m\beta\beta^* \leqslant \beta A\beta^* \leqslant M\,\beta\,\beta^*.$$

Therefore from inequality (20), taking into account formula (18), there follows

$$\beta\, B^{-1}\,\beta^* \geqslant m\,\beta\beta^*.$$

In a similar way, the second inequality in (17) can be shown to hold.

II. We now return to the investigation of the family of geodesics (13). It will be noted that to this family of geodesics there corresponds a set

Z_0 of values z lying inside and on the surface of the ellipsoid

$$zB^{-1}(0)\,z^* = 1, \tag{21}$$

because with each $z \in Z_0$ one can associate by formula (13) a geodesic crossing the hyperplane $y=0$. We find now what additional conditions the metric coefficients must satisfy for the family of geodesics to have in the region $\mathscr{D}\{0 \leqslant y \leqslant H\}$ a structure similar to that of the family of curves considered in Sec. 4. First of all it will be remembered that the family of curves in Sec. 4 had the following property: each curve of the family in $\mathscr{D}$ represented an arc with its endpoints on the hyperplane $y=0$ and the vertices of the family of curves filled out the whole region $\mathscr{D}$. We shall now study when the geodesics in metric (7) possess analogous properties.

Note the following fact: if at some point a geodesic has a vertex (a point most distant from the hyperplane $y=0$), then $|x_y|$ becomes infinite there. But one can see from (12) that this is possible only if the expression in the denominator Δ vanishes. Denoting the coordinates of the vertex of the geodesic by (ξ, η) we find

$$zB^{-1}(\eta)\,z^* = 1. \tag{22}$$

Let Z be a set of all z satisfying (22) when the parameter η varies over the range $0 \leqslant \eta \leqslant H$. From inequality (17) it immediately follows that the set Z lies in the annulus

$$0 < \frac{1}{\sqrt{M}} \leqslant |z| \leqslant \frac{1}{\sqrt{m}} < \infty .$$

For a geodesic to exist crossing the hyperplane $y=0$ and having a vertex in the hyperplane $y=\eta$, $0 \leqslant \eta \leqslant H$, it is necessary for the set Z satisfying (22) for fixed η to have a non-empty intersection with the set Z_0. Comparing (21) and (22) we see that the sets Z, Z_0 have at least a common set of points lying on the ellipsoid (21). It may however happen that they do not have other common points. For the family of geodesics to resemble in structure the family of curves in Sec. 4, it is necessary that $Z \subset Z_0$. We shall give a lemma which makes it possible to formulate a sufficient condition for Z to belong to Z_0 and at the same time to indicate the family of curves satisfying the above requirements.

Choose from the whole family of geodesics a family dependent on the $(n+1)$st parameter in the following way: take an arbitrary point $(x^0, 0)$ of the hyperplane $y=0$ and consider a one-parameter family of geodesics emanating from it and corresponding to the totality of $z \in Z_0$ which lie in $z_1, \ldots, z_n$ space on the ray $z = \alpha t$ (α is a fixed unit vector). Each such geodesic can be described by the parameters x^0, t; in the following, it will be denoted by $\Gamma_\alpha(x^0, t)$. In addition introduce the following notation:

$$v_\alpha(y) = \alpha B^{-1}(y)\,\alpha^*.$$

From the above assumptions about the coefficients a_{ij} it follows that $v_\alpha(y) \in C^2$. The following lemma holds.

Lemma 4. *For a geodesic $\Gamma_\alpha(x^0, t)$ with vertex in the hyperplane $y=\eta$ to exist for any $\eta \in [0, H]$, it is necessary and sufficient that the function $(d/dy)\,v_\alpha(y) = v'_\alpha(y)$ be positive in the interval $0 \leqslant y \leqslant H$.*

Proof. Suppose that for any $\eta \in [0, H]$ there exists a geodesic $\Gamma_\alpha(x^0, t)$ with vertex in the hyperplane $y=\eta$. We show that in this case the function $v'_\alpha(y)$ is positive in the interval $[0, H]$. For one thing, the segment of the ray $z=\alpha t$ belonging to Z must by hypothesis belong to Z_0. Consider a fixed $\eta_0 \in [0, H]$ and let there correspond to it the value of the parameter $z=z_0$. Then the geodesic with parameter $z_0=\alpha t_0$ must have vertex at the point $\eta=\eta_0$ and for this as follows from (13′), (11′), it is necessary that in a neighbourhood of the point $y=\eta_0$ the function

$$1 - zB^{-1}(y)\,z^* \equiv 1 - t_0^2\, v_\alpha(y)$$

decrease monotonically vanishing at the point itself. This requirement reduces to the fact that $v_\alpha(y)$ increase monotonically in a neighbourhood of $y=\eta_0$. Moreover, its derivative $v'_\alpha(y)$ at the point $y=\eta_0$ itself cannot vanish because in this case the integral (13′) would diverge as $y \to \eta_0$ which corresponds to the asymptotic approximation of the curve to the hyperplane $y=\eta_0$. Thus $v'_\alpha(\eta_0) > 0$. Since η_0 is an arbitrary point of $[0, H]$ it hence follows that $v'_\alpha(y)$ is positive in $[0, H]$.

We now prove the second half of the statement of the lemma. Namely, let $v'_\alpha(y)$ be positive in $[0, H]$; then for any $\eta_0 \in [0, H]$ there exists a geodesic $\Gamma_\alpha(x^0, t)$ with vertex in plane $y=\eta_0$. We first note that in this case the necessary condition for the existence of the geodesic $\Gamma_\alpha(x^0, t_0)$ with vertex in $y=\eta_0$ is fulfilled. Indeed, the segment of $z=\alpha t$ belonging to the region Z, also belongs to Z_0, since in this case the range of the parameter t for $z \in Z$ is included, as follows from (22), within $0 \leqslant v_\alpha^{-1/2}(H) \leqslant t \leqslant v_\alpha^{-1/2}(0)$ and the possible range of the parameter t for $z \in Z_0$ is included within $0 \leqslant t \leqslant v_\alpha^{-1/2}(0)$. However the positiveness of the derivative $v'_\alpha(y)$ turns out to be sufficient for the existence of the geodesic with vertex belonging to $\mathcal{D}$. To see this, consider an arbitrary value of $\eta_0 \in [0, H]$ and let there correspond to it the value $t_0 = v_\alpha^{-1/2}(\eta_0)$ computed from (22). Let us investigate the equation for the geodesic (13′) with the parameter $z=z_0=\alpha t_0$. As y increases from 0 to η_0, the expression in the denominator of Δ increases monotonically and at $y=\eta_0$ it vanishes. The further increase of y yields negative numbers under the square root.

This shows that for the geodesic to be continuable, it is necessary to decrease y from η_0 to zero. Thus the point $y=\eta_0$ is a vertex of the geodesic. Hence the statement is proved.

We point out now some further characteristics of the relation in this case between the parameters η and t. Let η_1, η_2 be arbitrary numbers in $[0, H]$ with $\eta_1<\eta_2$. Then $v_\alpha(\eta_1)<v_\alpha(\eta_2)$ and from (22), $t_1>t_2$. Thus to the points on $z=\alpha t$ lying nearer to the origin there correspond geodesics with vertices farther distant from $y=0$.

Corollary. *For each family $\Gamma_\alpha(x^0, t)$ to contain a subfamily of curves whose vertices fill out the region $\mathscr{D}$, it is necessary that for any $\eta_1, \eta_2 \in [0, H]$ such that $\eta_1<\eta_2$, the ellipsoid* (22), *corresponding to the value $\eta=\eta_2$, lies exactly inside the ellipsoid* (22) *corresponding to the value $\eta=\eta_1$.*

III. Let α be a fixed unit vector and let the coefficients of metric (7) be such that $v'_\alpha(y)>0$ in $[0, H]$. Consider on the family $\Gamma_\alpha(x^0, t)$ the integral-geometric problem for the function $u(x, y)$:

$$\int_{\Gamma_\alpha(x^0, t)} u(x, y)\, ds = v_\alpha(x^0, t). \tag{23}$$

$$v_\alpha^{-1/2}(H) \leqslant t \leqslant v_\alpha^{-1/2}(0),\; x^0 \in \{y=0\}.$$

In this case, ds is naturally assumed to be the element of arclength in the Riemannian metric (7). Let us show that (23) has a unique solution. In this connection we shall not try to show that all conditions 1°–5° of Sec. 4 hold for $\Gamma_\alpha(x^0, t)$, but we shall apply the method of Sec. 4 directly. Note that the equation of curves $\Gamma_\alpha(x^0, t)$ for $v_\alpha^{-1/2}(H) \leqslant t \leqslant v_\alpha^{-1/2}(0)$ can be represented in a more customary form for the method of Sec. 4. Namely, (13′) can be written as

$$x = \xi - \int_y^\eta \left[\frac{q(y)}{s(y)} + (-1)^{j+1} \frac{\alpha B^{-1}(y)}{\sqrt{s(y)\,[v_\alpha(\eta) - v_\alpha(y)]}}\right] dy, \tag{24}$$

$$(0 \leqslant y \leqslant \eta \leqslant H), \qquad (j=1, 2)$$

where the parameters η and t are related to each other by the one-to-one relation

$$v_\alpha(\eta) = \frac{1}{t^2}, \tag{25}$$

and ξ is given by the formula

$$\xi = x^0 + \int_0^\eta \left[\frac{q(y)}{s(y)} + \frac{\alpha\, B^{-1}(y)}{\sqrt{s(y)\,[v_\alpha(\eta) - v_\alpha(y)]}}\right] dy. \tag{26}$$

From (24) it follows that the condition

$$0 \leqslant k_0 \leqslant v'_\alpha(y) \leqslant k_1 < \infty, \ (0 \leqslant y \leqslant H) \tag{27}$$

is sufficient for the curves $\Gamma_\alpha(x^0, t)$ to be of the same type in a neighbourhood of the vertex as the curves (4.14). But the above condition is fulfilled in this case by virtue of the assumption of positiveness of the continuous function $v'_\alpha(y)$ on the closed interval $[0, H]$.

Taking into account (8), (11'), (25), we can write equation (23) in the form

$$\sum_{j=1}^{2} \int_0^\eta u(x_j, y) \frac{\sqrt{v_\alpha(\eta)}}{\sqrt{s(y)\,[v_\alpha(\eta) - v_\alpha(y)]}}\, dy = v_\alpha(x^0, v_\alpha^{-1/2}(\eta)). \tag{28}$$

In this equation x_j is given by formula (24). Taking Fourier transforms in (28) in the variable ξ related to x^0 by (26) we arrive at the following equation for the Fourier transform $u_\lambda(y)$ of $u(x, y)$:

$$\int_0^\eta u_\lambda(y)\, K_\lambda(y, \eta)\, dy = v_{\alpha\lambda}(\eta), \tag{28'}$$

where the kernel is given by

$$K_\lambda(y, \eta) = \frac{2\sqrt{v_\alpha(\eta)}}{\sqrt{s(y)\,[v_\alpha(\eta) - v_\alpha(y)]}} \cos\left\{\int_y^\eta \frac{(\lambda, \alpha B^{-1}(y))\, dy}{\sqrt{s(y)\,[v_\alpha(\eta) - v_\alpha(y)]}}\right\} \cdot \exp\left\{i \int_y^\eta \frac{(\lambda, q(y))}{s(y)}\, dy\right\}. \tag{29}$$

The properties of the kernel of (28') resemble those of (4.12) and the kernel of (28') is, in a sense, even better than that of (4.12). From (29) and the differentiability properties of the functions in it, it is clear that the function $K_\lambda(y, \eta)$ can be represented in the form

$$K_\lambda(y, \eta) = \frac{1}{\sqrt{\eta - y}} R_\lambda(y, \eta), \tag{30}$$

where $R_\lambda(y, \eta)$ is a continuously differentiable function in the region $\{0 \leqslant y \leqslant \eta \leqslant H\}$ such that

$$R_\lambda(\eta, \eta) = 2\sqrt{\frac{v_\alpha(\eta)}{s(\eta)\, v'_\alpha(\eta)}} \geqslant 2\sqrt{\frac{m}{mk_1}} > 0. \tag{31}$$

Because of these properties of the kernel the application of the operator L of (4.14) to (28′) immediately yields a Volterra equation of the second kind with continuous kernel. Thus we have proved the following theorem.

Theorem 1.11. *Let* $u(x, y)$ *be a continuous finite function in the region* $\mathscr{D}\{0 \leqslant y \leqslant H\}$ *and let* $\Gamma_\alpha(x^0, t)$ *be a family of geodesics in metric* (7) *whose coefficients are twice continuously differentiable functions satisfying condition* (27). *Then* $u(x, y)$ *is uniquely determined by integrals* (23).

Analogously to Sec. 4 one can introduce in (23) a known weight function $\varrho(x - x^0, y, t)$ which is continuously differentiable in all arguments and different from zero at the vertex of each curve $\Gamma_\alpha(x^0, t)$. Again, the uniqueness theorem remains valid.

Chapter II

Inverse Problems for Hyperbolic Linear Differential Equations

1. General Information Concerning the Solution of the Cauchy Problem for Linear Hyperbolic Equations

In this chapter we shall consider second order linear differential equations of hyperbolic type and problems involving point sources. Such problems are of the type

$$\left.\begin{aligned} u_{tt} &= \mathrm{L}u + 4\,\pi\delta(x-x^0,\, t), \\ u|_{t<0} &\equiv 0. \end{aligned}\right\} \qquad (1)$$

Here $u(x; t)$ is the unknown function, $x^0=(x_1^0,\dots, x_n^0)$ is a fixed point and $x=(x_1,\dots, x_n)$ a variable point of $x_1,\dots, x_n$ space; $\delta(x-x^0, t)$ is Dirac's delta function [27] and L the elliptic differential operator

$$\left.\begin{aligned} \mathrm{L}u &\equiv \sum_{i,j=1}^{n} a_{ij}(x)\, u_{x_i x_j} + \sum_{i=1}^{n} b_i(x)\, u_{x_i} + c(x)\, u, \\ &\sum_{i,j=1}^{n} a_{ij}(x)\, t_i t_j \geqslant \mu_0 \sum_{i=1}^{n} t_i^2 \quad (\mu_0>0). \end{aligned}\right\} \qquad (2)$$

If problem (1) is considered in unbounded space, then boundary conditions may be specified for $u(x, t)$ in addition to the initial conditions.

The problem of determining $u(x, t)$ satisfying (1) with known coefficients a_{ij}, b_i, c is the so-called direct problem for the differential equation. The main purpose of our further investigations is the formulation and analysis of problems inverse to the one mentioned above. Namely, we shall be interested in formulations of the sort wherein it is possible to find, from known functionals of the solution of problem (1), all or a portion of the coefficients of the differential equation. In practical applications most important are the problems requiring a minimal number of functionals to uniquely determine the coefficients of the equation. It is clear that in general this minimal information about the solution of a problem must be of the same dimension as the information about the coefficients which are being determined. The present chapter contains several versions of problems with a minimal amount of the information used.

In investigating problems of type (1) we shall use the procedure of S. L. Sobolev. In [72], [73] S. L. Sobolev devised a method for solving the Cauchy problem for the three-dimensional wave equation with variable coefficient. Later V. G. Gogoladzé, on the basis of S. L. Sobolev's method, extended the results on the Cauchy problem (see [30]–[32]) to the "generalized wave equation" (terminology introduced by V. G. Gogoladzé):

$$u_{tt}=\sum_{i=1}^{3}(a_i u_{x_i x_i}+b u_{x_i})+cu.$$

Then in [76] S. L. Sobolev extended the above method to general second order linear hyperbolic equations.

We shall present now some results for problem (1) due to S. L. Sobolev. For simplicity we shall consider only the case $n=3$. First we shall discuss a more general Cauchy problem than (1), namely,

$$\left.\begin{aligned} &u_{tt}=\sum_{i,j=1}^{3}a_{ij}u_{x_i x_j}+\sum_{i=1}^{3}b_i u_{x_i}+cu+f(x,t),\\ &u(x,0)=\varphi(x),\qquad u_t(x,0)=\psi(x).\end{aligned}\right\}\tag{3}$$

Let the coefficients $a_{ij}(x)$ be sufficiently smooth and such that to each pair of points x, $\xi(\xi=(\xi_1,\xi_2,\xi_3))$ there corresponds a single geodesic $\Gamma(x,\xi)$ joining them in the metric in which the element of length ds is given by

$$ds=\sqrt{\sum_{i,j=1}^{3}b_{ij}(x)\,dx_i\,dx_j},\tag{4}$$

and the b_{ij} are the elements of the matrix inverse to $A=\|a_{ij}\|$. Let $\tau(x,\xi)$ denote the distance in metric (4) between the points x and ξ. In other words,

$$\tau(x,\xi)=\int_{\Gamma(x,\xi)}\sqrt{\sum_{i,j=1}^{3}b_{ij}\,dx_i\,dx_j}.\tag{5}$$

Physically, $\tau(x,\xi)$ is the propagation time for a disturbance produced at point ξ to reach point x. Evidently, $\tau(x,\xi)=\tau(\xi,x)$. S. L. Sobolev ascertained the existence of a function $\sigma(x,\xi)$ satisfying in its first argument the equation

$$2\sum_{i,j=1}^{3}a_{ij}\tau_{x_i}\sigma_{x_j}+\sigma\left[\sum_{i,j=1}^{3}a_{ij}\tau_{x_i x_j}+\sum_{i=1}^{3}\left(2\sum_{j=1}^{3}\frac{\partial a_{ij}}{\partial x_j}-b_i\right)\tau_{x_i}\right]=0,\tag{6}$$

and possessing a number of remarkable properties (see below) owing to which problem (3) reduces to the integral equation

$$
\begin{aligned}
u(x,t) = \frac{1}{4\pi} \iint\limits_{S(x,t)} & \{\sigma(\xi,x)\,\mathrm{N}_\xi\varphi(\xi) - \varphi(\xi)\,\mathrm{N}_\xi\sigma(\xi,x) \\
& + \sigma(\xi,x)\,[\psi(\xi)\,\mathrm{N}_\xi\tau(\xi,x) + \varphi(\xi)\,R_\xi]\}\,dS_\xi \\
& + \frac{1}{4\pi} \iiint\limits_{\mathcal{D}(x,t)} \sigma(\xi,x)\,f(\xi, t-\tau(\xi,x))\,dv_\xi \\
& + \frac{1}{4\pi} \iiint\limits_{\mathcal{D}(x,t)} \mathrm{L}_\xi^*\sigma(\xi,x)\cdot u(\xi, t-\tau(\xi,x))\,dv_\xi .
\end{aligned}
\tag{7}
$$

Here ξ is the variable point of integration, $S(x, t)$ a quasisphere $\tau(x, \xi)=t$ and $\mathcal{D}(x, t)$ the volume bounded by this quasisphere; ds and dv are the elements of surface area and volume respectively; L^* is the operator adjoint to L in the sense of Lagrange. N is the linear differential operator

$$\mathrm{N}\varphi(x) = \sum_{i,j=1}^{3} a_{ij}\varphi_{x_i}\cos(n, x_i);$$

the subscript ξ on the operators L^* and N shows that they are to be applied to the variable ξ;

$$R = \sum_{i=1}^{3}\left(b_i - \sum_{j=1}^{3}\frac{\partial a_{ij}}{\partial x_j}\right)\cos(n, x_i);$$

n is the unit external normal to the surface $S(x, t)$.

Note that as x and t vary, the region of integration $\mathcal{D}(x, t)$ also varies and so (7) is analogous to a Volterra equation. S. L. Sobolev showed that (7) has a unique solution in that region where the geodesics $\Gamma(x, \xi)$ behave regularly and this solution can be obtained by the method of successive approximations. Moreover, the solution of (7) thus obtained is also the solution of problem (3).

We shall indicate now the basic properties of the function $\sigma(x, \xi)$. As mentioned above, it satisfies the first order partial differential equation (6). In addition, the function $\sigma(x, \xi)$ is smooth in a neighbourhood of the point ξ (its smoothness, naturally, depends on that of the coefficients of the operator L) and at ξ itself it has singularity and satisfies the following conditions:

1) The product $\sigma(x, \xi)\,\tau(x, \xi)$ is a smooth function, including the

point ξ, and

$$\lim_{x\to\xi} \sigma(x, \xi)\, \tau(x, \xi) = \frac{1}{\lambda_1(\xi)\, \lambda_2(\xi)\, \lambda_3(\xi)}, \tag{8}$$

where the $\lambda_i(\xi)$ $(i=1, 2, 3)$ are the semi-axes of the ellipsoid

$$\sum_{i,j=1}^{3} b_{ij}(\xi)\, x_i x_j = 1. \tag{9}$$

2) The operator L^* applied to $\sigma(x, \xi)$ satisfies the inequality

$$|L_x^* \sigma(x, \xi)| \leqslant \frac{K}{\tau(x, \xi)}, \tag{10}$$

where K is a positive constant dependent only on the global properties of the coefficients of the operator L^*.

3) If S_1 is a closed surface, containing the point ξ inside and n is the direction of the external normal on S_1, then on shrinking the surface S_1 to the point ξ there results the limiting relation

$$\lim_{S_1 \to \xi} \iint_{S_1} N_x \sigma(x, \xi)\, dS = -4\pi. \tag{11}$$

If $a_{ij} = \delta_{ij}$ (δ_{ij} is the Kronecker delta) and $b_i = 0$, the function $\sigma(x, \xi) = \frac{1}{|x-\xi|}$ satisfies all the above conditions. If the geodesics in metric (4) are regarded as known, then the function $\sigma(x, \xi)$ satisfying (6) and possessing the above properties, can be easily found. It has the form

$$\sigma^2(x, \xi) = \frac{\sin\theta \exp\left\{ \int_{\Gamma(x,\xi)} \sum_{i=1}^{3} \left[\left(b_i - \sum_{j=1}^{3} \frac{\partial a_{ij}}{\partial x_j} \right) \sum_{k=1}^{3} b_{ik} \frac{dx_k}{ds} \right] ds \right\}}{\sum_{i,j=1}^{3} b_{ij}(\xi)\, \alpha_i \alpha_j \left| \frac{\partial(x_1, x_2, x_3)}{\partial(s, \theta, \varphi)} \right| \lambda_1(\xi)\, \lambda_2(\xi)\, \lambda_3(\xi)}, \tag{12}$$

$$(\alpha_1 = \sin\theta \cos\varphi, \quad \alpha_2 = \sin\theta \sin\varphi, \quad \alpha_3 = \cos\theta)$$

where s, θ, φ are curvilinear coordinates defined as follows: θ and φ are the angular coordinates in a spherical coordinate system corresponding to the direction of the tangent to the geodesic $\Gamma(x, \xi)$ at ξ and s is the arclength along the geodesic $\Gamma(x, \xi)$ in metric (4); $\frac{\partial(x_1, x_2, x_3)}{\partial(s, \theta, \varphi)}$ is the Jacobian of the transformation from Cartesian coordinates at point x to the curvilinear coordinates.

We now return to problem (1). Eq. (7) thus takes the form

$$u(x,t)=\sigma(x^0,x)\,\delta(t-\tau(x^0,x)) + \frac{1}{4\pi}\iiint\limits_{\mathcal{D}(x,t)} \mathrm{L}_\xi^*\sigma(\xi,x)\,u(\xi,t-\tau(\xi,x))\,dv_\xi. \tag{13}$$

In the above formula the region of integration actually degenerates into the region $\mathcal{D}(x,x^0,t)$ bounded by the surface of the quasi-ellipsoid

$$S(x,x^0,t)\colon \{\tau(\xi,x)+\tau(\xi,x^0)=t\},$$

since the only points x, t affecting the value of the function $u(x,t)$ at point ξ are those for which the sum of propagation times of disturbances from the point x^0 to ξ and from ξ to x does not exceed t. Introducing a new function $v(x,t)$ defined by

$$u(x,t)=v(x,t)+\sigma(x^0,x)\,\delta(t-\tau(x^0,x)), \tag{14}$$

we obtain the equation

$$\begin{aligned} v(x,t)=&\frac{1}{4\pi}\iiint\limits_{\mathcal{D}(x,x^0,t)} \mathrm{L}_\xi^*\sigma(\xi,x)\,\sigma(x^0,\xi)\,\delta(t-\tau(\xi,x)-\tau(\xi,x^0))\,dv_\xi \\ &+\frac{1}{4\pi}\iiint\limits_{\mathcal{D}(x,x^0,t)} \mathrm{L}_\xi^*\sigma(\xi,x)\cdot v(\xi,t-\tau(\xi,x))\,dv_\xi. \end{aligned} \tag{15}$$

Let us show that $v(x,t)$ is continuous in x and t in the region

$$\mathcal{G}(x^1,x^0,T)\colon \{\tau(x^0,x)\leqslant t\leqslant T-\tau(x^1,x)\},$$

and that it can be found from (15) by the method of successive approximations. Here x^1 is an arbitrary fixed point in three-dimensional space and T is an arbitrary positive number satisfying the condition

$$T\geqslant\tau(x^1,x^0).$$

To this end, we introduce curvilinear coordinates for the point ξ. Let ξ lie in the quasi-ellipsoid

$$S(x,x^0,t_1)\colon \{\tau(\xi,x)+\tau(\xi,x^0)=t_1\}.$$

Consider the geodesic $\Gamma(x^0,\xi)$ and denote by τ the travel time of a signal along this geodesic from x^0 to ξ, i.e. $\tau=\tau(x^0,\xi)$. t_1 and τ will be regarded as curvilinear coordinates of ξ. As the third curvilinear coordinate, we take the angle φ between the two planes, one of which is fixed and contains the tangent dx/ds to the geodesic $\Gamma(x,x^0)$ at the point x^0 and the second of which passes through x^0 and contains at this point the tangents to the curves $\Gamma(x^0, \xi)$ and $\Gamma(x^0, x)$. The angle φ here is an angle related to metric (4). The cosine of the angle between vectors α and β is introduced in the usual way through the scalar product by

$$\cos(\alpha, \beta) = \frac{(\alpha, \beta)}{\sqrt{(\alpha, \alpha)(\beta, \beta)}},$$

and the scalar product (α, β) of the vectors $\alpha = (\alpha_1, \ldots, \alpha_n)$, $\beta = (\beta_1, \ldots, \beta_n)$, in turn, is defined by (see [63])

$$(\alpha, \beta) = \sum_{i,j=1}^{n} b_{ij} \alpha_i \beta_j.$$

The curvilinear coordinates t_1, τ, φ determine ξ uniquely. This follows from the one-to-one correspondence between each pair of points x and ξ and the geodesic $\Gamma(x, \xi)$ joining them. Changing in (15) from the integration variables ξ_1, ξ_2, ξ_3 to t_1, τ, φ, we find

$$\begin{aligned} v(x, t) = {} & \frac{1}{4\pi} \iint\limits_{S(x, x^0, t)} L_\xi^* \sigma(\xi, x) \cdot \sigma(x^0, \xi) \left| \frac{\partial(\xi_1, \xi_2, \xi_3)}{\partial(t, \tau, \varphi)} \right| d\tau \, d\varphi \\ & + \frac{1}{4\pi} \int\limits_{\tau(x, x^0)}^{t} dt_1 \iint\limits_{S(x, x^0, t_1)} L_\xi^* \sigma(\xi, x) \\ & \times \left| \frac{\partial(\xi_1, \xi_2, \xi_3)}{\partial(t_1, \tau, \xi)} \right| v(\xi, t - \tau(\xi, x)) \, d\tau \, d\varphi, \end{aligned} \tag{16}$$

where $\dfrac{\partial(\xi_1, \xi_2, \xi_3)}{\partial(t_1, \tau, \varphi)}$ is the Jacobian of the transformation from Cartesian to curvilinear coordinates. In what follows we shall need an estimate for the Jacobian. However, precisely the same Jacobian was considered by S. L. Sobolev in his paper [73] on the integration of the wave equation in an inhomogeneous medium. In this work S. L. Sobolev introduced bipolar-geodesic coordinates τ_1, τ, φ for ξ, where $\tau_1 = \tau(\xi, x)$, and he estimated the Jacobian of the transformation from Cartesian coordinates to bipolar-geodesic coordinates. Since our coordinates differ from bipolar-geodesic coordinates merely in that we use $t_1 = \tau + \tau_1$ instead of τ_1, the Jacobians of these two transformations of the coordinate systems coincide. S. L. Sobolev established under certain conditions on the surfaces $\tau(\xi, x^0) = t$, $\tau(\xi, x) = t$, the following estimate:

$$\left| \frac{\partial(\xi_1, \xi_2, \xi_3)}{\partial(\tau_1, \tau, \varphi)} \right| \leqslant N_1 \tau_1 \tau, \tag{17}$$

where N_1 is constant. One of these conditions is as follows. Consider an arbitrary point ξ of the extremal $\Gamma(x, x^0)$ and pass two tangent surfaces $\tau(\xi, x^0)=\text{const.}$ and $\tau(\xi, x)=\text{const.}$ through this point. The extremal $\Gamma(x, x^0)$ is to be orthogonal, in the sense of metric (4), to both surfaces at ξ. Consider now the curvatures of the normal cross-sections of both surfaces, i.e. the cross-sections in the planes passing through the tangent to the extremal at ξ. The curvature κ of the cross-section will be considered positive if its center lies on the normal in the direction of x^0. Then the condition amounts to the requirement that for any two normal cross-sections of the surfaces $\tau(\xi, x^0)=\text{const.}$, $\tau(\xi, x)=\text{const.}$ the difference of their respective curvatures κ, κ_1 in absolute value should have a greatest lower bound different from zero,

$$|\kappa-\kappa_1|\geqslant m>0.$$

This is a natural condition if the points x and x^0 do not come arbitrarily close. In the latter case it is easy to verify, taking $b_{ij}=\text{const.}$, that a more natural condition is

$$|\kappa-\kappa_1|\geqslant m/\tau(x, x^0), \qquad (m>0)$$

which we shall assume. But then, analyzing the derivation of formula (7), we easily observe that the constant N_1 must increase inversely proportionally to $\tau(x, x^0)$; and hence, instead of (17), the formula

$$\left|\frac{\partial(\xi_1, \xi_2, \xi_3)}{\partial(\tau_1, \tau, \varphi)}\right|=\left|\frac{\partial(\xi_1, \xi_2, \xi_3)}{\partial(t_1, \tau, \varphi)}\right|\leqslant N\frac{\tau\ \tau_1}{\tau(x, x^0)} \tag{18}$$

must hold, in which N is a positive constant depending on the general properties of the coefficients b_{ij}. When the b_{ij} are constant, the equality sign holds in formula (18).

Let us write the double integral in (16) over the surface $S(x, x^0, t)$ in the form of an iterated integral. For this purpose, we shall determine the range of values of τ in this integral. We show that the points ξ^1 and ξ^2 at which τ attains its minimum and maximum values, lie on the extensions of the geodesic $\Gamma(x, x^0)$. Indeed, let ξ^1 be the point of intersection of the quasi-ellipsoid $S(x, x^0, t_1)$ with the geodesic $\Gamma(x, x^0)$, where the arrangement of the points ξ_1, x^0, x is such that x^0 lies between ξ^1 and x. Let ξ be an arbitrary point of the surface $S(x, x^0, t)$. Then

$$\left.\begin{aligned}2\tau(\xi^1, x^0)+\tau(x^0, x)&=t_1,\\ \tau(\xi, x)+\tau(\xi, x^0)&=t_1.\end{aligned}\right\} \tag{19}$$

Subtracting the second equation from the first and using the triangle inequality, we obtain

$$2[\tau(\xi^1, x^0)-\tau(\xi, x^0)]=\tau(\xi, x)-\tau(\xi, x^0)-\tau(x^0, x)\leqslant 0.$$

Thus $\tau=\tau(\xi^1, x^0)$ is the minimum value of τ. From the first formula of (19) we find

$$\tau_{\min}=\tfrac{1}{2}[t_1-\tau(x, x^0)].$$

In a similar way, it is proved that the second point of intersection of the continuation of the geodesic $\Gamma(x, x^0)$ with the surface $S(x, x^0, t_1)$ yields the maximum value equal to

$$\tau_{\max}=\tfrac{1}{2}[t_1+\tau(x, x^0)].$$

Eq. (16) can then be written as

$$\begin{aligned} v(x, t)=&\frac{1}{4\pi}\int_0^{2\pi} d\varphi \int_{\frac{1}{2}[t-\tau(x, x^0)]}^{\frac{1}{2}[t+\tau(x, x^0)]} \mathrm{L}_\xi^*\sigma(\xi, x)\,\sigma(x^0, \xi)\left|\frac{\partial(\xi_1, \xi_2, \xi_3)}{\partial(t, \tau, \varphi)}\right| d\tau \\ &+\frac{1}{4\pi}\int_{\tau(x, x^0)}^{t} dt_1 \int_0^{2\pi} d\varphi \int_{\frac{1}{2}[t_1-\tau(x, x^0)]}^{\frac{1}{2}[t_1+\tau(x, x^0)]} \mathrm{L}_\xi^*\sigma(\xi, x)\left|\frac{\partial(\xi_1, \xi_2, \xi_3)}{\partial(t_1, \tau, \varphi)}\right| \\ &\cdot v(\xi, t-\tau(\xi, x))\, d\tau. \end{aligned} \tag{20}$$

We apply the method of successive approximations to (20) according to the scheme

$$\begin{aligned} v_0(x, t)=&\frac{1}{4\pi}\int_0^{2\pi} d\varphi \int_{\frac{1}{2}[t-\tau(x, x^0)]}^{\frac{1}{2}[t+\tau(x, x^0)]} \mathrm{L}_\xi^*\sigma(\xi, x)\,\sigma(x^0, \xi)\left|\frac{\partial(\xi_1, \xi_2, \xi_3)}{\partial(t, \tau, \varphi)}\right| d\tau, \\ v_n(x, t)=&\,v_0(x, t) \\ &+\frac{1}{4\pi}\int_{\tau(x, x^0)}^{t} dt_1 \int_0^{2\pi} d\varphi \int_{\frac{1}{2}[t_1-\tau(x, x^0)]}^{\frac{1}{2}[t_1+\tau(x, x^0)]} \mathrm{L}_\xi^*\sigma(\xi, x)\left|\frac{\partial(\xi_1, \xi_2, \xi_3)}{\partial(t_1, \tau, \varphi)}\right| \\ &\cdot v_{n-1}(\xi, t-\tau(\xi, x))\, d\tau, \qquad (n=1, 2, 3, \ldots) \end{aligned} \tag{21}$$

and we estimate the successive approximations $v_n(x, t)$. If we use formulas (8), (10), (18), we find

$$\begin{aligned} &|v_0(x, t)| \\ &\leqslant\frac{1}{4\pi}\int_0^{2\pi} d\varphi \int_{\frac{1}{2}[t-\tau(x, x^0)]}^{\frac{1}{2}[t+\tau(x, x^0)]} \frac{K}{\tau(\xi, x)}\,\sigma(x^0, \xi)\,\frac{N\ \tau\ \tau(\xi, x)}{\tau(x, x^0)}\, d\tau \end{aligned}$$

$$\leqslant \frac{R}{\tau(x,x^0)} \int\limits_{\frac{1}{2}[t-\tau(x,x^0)]}^{\frac{1}{2}[t+\tau(x,x^0)]} d\tau = R,$$

where

$$R = \tfrac{1}{2}KN \max_{\xi \in \mathscr{G}(x^1, x^0, T)} [\tau(x^0, \xi)\,\sigma(x^0, \xi)].$$

The next approximation is estimated in a similar way to yield

$$|v_1(x,t)| \leqslant R + \frac{1}{4\pi} \int\limits_{\tau(x,x^0)}^{t} dt_1 \int\limits_0^{2\pi} d\varphi \int\limits_{\frac{1}{2}[t_1-\tau(x,x^0)]}^{\frac{1}{2}(t_1+\tau(x,x^0)]} \frac{K}{\tau(\xi,x)} \frac{N\ \tau\ \tau(\xi,x)}{\tau(x,x^0)} \cdot R\,d\tau$$
$$= R\left[1 + \frac{KN}{(1!)^2 \cdot 2} \frac{t^2-\tau^2(x,x^0)}{4}\right].$$

We now apply mathematical induction to show that the n-th approximation satisfies the following estimate:

$$|v_n(x,t)| \leqslant R \sum_{m=0}^{n} \frac{(KN)^m}{(m!)^2\,(m+1)} \left[\frac{t^2-\tau^2(x,x^0)}{4}\right]^m. \tag{22}$$

For $n=0, 1$ the estimate is valid. Assume that formula (22) holds for the n-th approximation and let us show that it also holds for the $(n+1)$st approximation. Indeed,

$$|v_{n+1}(x,t)| \leqslant R + \frac{1}{4\pi} \int\limits_{\tau(x,x^0)}^{t} dt_1 \cdot \int\limits_0^{2\pi} d\varphi \int\limits_{\frac{1}{2}[t_1-\tau(x,x^0)]}^{\frac{1}{2}[t_1+\tau(x,x^0)]} \frac{K}{\tau(\xi,x)} \cdot \frac{N\cdot\tau\cdot\tau(\xi,x)}{\tau(x,x^0)}$$
$$\cdot R \sum_{m=0}^{n} \frac{(KN)^m}{(m!)^2\,(m+1)} \left[\frac{(t-t_1+\tau)^2-\tau^2}{4}\right]^k d\tau$$
$$= R\left[1 + \sum_{m=0}^{n} \frac{(KN)^{m+1}}{(m!)^2\,(m+1)} \frac{1}{2\tau(x,x^0)} \int\limits_{\tau(x,x^0)}^{t} dt_1 \int\limits_{\frac{1}{2}[t_1-\tau(x,x^0)]}^{\frac{1}{2}[t_1+\tau(x,x^0)]} \left(\frac{t-t_1}{2}\right)^k\right.$$
$$\left.\cdot\left(\tau + \frac{t-t_1}{2}\right)^k \tau\,d\tau = R\left\{1 + \sum_{m=0}^{n} \frac{(KN)^{m+1}}{[(m+1)!]^2\,(m+2)} \left[\frac{t^2-\tau^2(x,x^0)}{4}\right]^{m+1}\right\}\right.$$
$$= R \sum_{m=0}^{n+1} \frac{(KN)^m}{(m!)^2\,(m+1)} \left[\frac{t^2-\tau^2(x,x^0)}{4}\right]^m.$$

Thus the estimate (22) holds for any n. Evidently, each function $v_n(x, t)$ is continuous in the region $\mathscr{G}(x^1, x^0, T)$. Moreover, the inequality

$$t^2-\tau^2(x, x^0)\leqslant T^2-\tau^2(x^1, x^0) \tag{23}$$

holds in this region for any x and t (this inequality will be derived a little later) and hence the series

$$\sum_{m=0}^{\infty}\frac{(KN)^m}{(m!)^2(m+1)}\left[\frac{t^2-\tau^2(x, x^0)}{4}\right]^m$$

is majorized by the convergent series

$$\sum_{m=0}^{\infty}\frac{(KN)^m}{(m!)^2(m+1)}\left[\frac{T^2-\tau^2(x^1, x^0)}{4}\right]^m.$$

Because of this the sequence of functions $v_n(x, t)$ converges uniformly in the region $\mathscr{G}(x^1, x^0, T)$ to a continuous function $v(x, t)$ yielding the unique solution of (20).

Let us show now the validity of inequality (23). According to definition the region $\mathscr{G}(x^1, x^0, T)$ is the set of points satisfying the inequality

$$\tau(x, x^0)\leqslant t\leqslant T-\tau(x, x^1).$$

If we square both sides of the inequality and subtract $\tau^2(x, x^0)$ we find

$$\begin{aligned}0\leqslant t^2-\tau^2(x, x^0)\leqslant{}& T^2-\tau^2(x^1, x^0)\\ &+[\tau^2(x^1, x^0)+\tau^2(x, x^1)-2T\tau(x, x^1)-\tau^2(x, x^0)].\end{aligned}$$

To be convinced of the validity of inequality (23), it suffices to prove that the expression in brackets is non-positive. Remembering that $T\geqslant\tau(x^1, x^0)$ and using the well-known inequality for the sides of a curvilinear triangle with vertices at the points x^0, x^1 and x, we obtain

$$\begin{aligned}&\tau^2(x^1, x^0)+\tau^2(x, x^1)-2T\tau(x, x^1)-\tau^2(x, x^0)\\ &\quad\leqslant\tau^2(x^1, x^0)+\tau^2(x, x^1)-2\tau(x^1, x^0)\,\tau(x, x^1)-\tau^2(x, x^0)\\ &\quad\leqslant[\tau(x^1, x^0)-\tau(x, x^1)]^2-\tau^2(x, x^0)]\leqslant 0.\end{aligned}$$

Hence, inequality (19) holds in the region $\mathscr{G}(x^1, x^0, T)$.

2. One-Dimensional Inverse Problem for the Telegraph Equation in Three-Dimensional Space

Consider in the region $y\geqslant 0$ the equation

$$u_{tt}=\Delta u+a(M)\,u+4\pi\delta(M-M_0, t) \tag{1}$$

under the initial and boundary conditions

$$u(M, M_0, t)|_{t<0} \equiv 0, \quad u_y(M, M_0, t)|_{y=0} = 0, \tag{2}$$

where $M(x_1, x_2, y)$ and $M_0(x_1^0, x_2^0, 0)$ are points in three-dimensional space, Δ is the Laplacian in the variables $x, y; x=(x_1, x_2)$ and $\delta(M-M_0, t)$ is Dirac's delta function. We pose the problem: Determine the coefficient $a(M)$ of (1) knowing the generalized solution of this equation in the plane $y=0$. Such a problem for (1) was first considered by Ju. M. Berezanskiĭ [9]. He proved a uniqueness theorem for the determination of the coefficient $a(M)$ in the class of piecewise-analytic functions in terms of the function $u(M, M_0, t)$ in which the points M and M_0 vary over a two-dimensional region of the plane $y=0$, i.e. in terms of a solution of (1) given as a function of five arguments.

In this section, we consider the problem of finding the coefficient $a(M)$ depending only on the coordinate y. Correspondingly less information is required to find such a function. Namely, it will be assumed that the solution of (1) is known under the conditions (2) only at one arbitrary point of the plane $y=0$. In this connection the point M_0 is assumed to be fixed.

The main result of this section is an existence and uniqueness theorem for the solution of the formulated problem in some sufficiently small interval in the class of continuous functions. The theorem is of a constructive character.

I. Since in our case the point M_0 does not vary, we shall for brevity write the solution of (1) in the form $u(M, t)$ instead of $u(M, M_0, t)$. In addition, it will be assumed that the origin is at M_0. For convenience we shall continue $a(M)$, $u(M, t)$ as even functions into the half-space $y<0$. Write the solution $u(M, t)$ in the form

$$u(M, t) = 2\frac{\delta(t-r(M, M_0))}{r(M, M_0)} + v(M, t). \tag{3}$$

Then using Kirchhoff's formula for the function $v(M, t)$, we obtain the integral equation

$$\begin{aligned} v(M, t) = &\frac{1}{2\pi}\iiint\limits_{r(P,M)\leqslant t} a(P)\frac{\delta(t-r(P, M_0)-r(P, M))}{r(P, M_0)\, r(P, M)}\, dv_P \\ &+\frac{1}{4\pi}\iiint\limits_{r(P,M)\leqslant t} a(P)\frac{v(P, t-r(P, M))}{r(P, M)}\, dv_P. \end{aligned} \tag{4}$$

In the above formulas $r(M, M_0)$ denotes Euclidean distance between M and M_0, dv_p is the volume element, and $P(\xi, \eta)$ $(\xi=(\xi_1, \xi_2))$ is the variable point of integration. It is easy to see that the region of integration in (4) actually degenerates to a region bounded by the ellipsoid of revolution

$$S(M, t): \{r(P, M) + r(P, M_0) \leqslant t\},$$

one of whose foci is at the point M_0 and the other at M. Indeed, only those points P affect the solution of (4) at point M at time t, for which the sum of the travel times of disturbances from M_0 to P and from P to M does not exceed t. If we take into account the fact that the propagation speed of disturbances in our case is equal to unity, the statement becomes obvious.

We introduce, along with the stationary ξ, η system coincident with the x, y system, a non-stationary Cartesian coordinate system ξ^1, η^1 ($\xi^1 = (\xi_1^1, \xi_2^1)$) connected to point M. We place its origin at M_0 and direct the η^1 axis along the straight line, joining the points M_0, M, towards M. We dispose the ξ_1^1, ξ_2^1 axes orthogonally to each other and to the η^1 axis so that one of them lies in the plane passing through the η and η^1 axes and so that the ξ^1, η^1 system is oriented in the same way as the ξ, η system. Moreover we consider spherical coordinates connected in the usual way with ξ, η and ξ^1, η^1. Let the position of the η^1 axis be determined with respect to the ξ, η system by the angles θ_0, φ_0 of the spherical coordinates, and the position of the radius vector M_0P with respect to the spherical system associated with ξ^1, η^1, by the angles θ, φ. Let $r = |M_0P|$. Then

$$\xi_1^1 = r \sin\theta \cos\varphi, \quad \xi_2^1 = r \sin\theta \sin\varphi, \quad \eta^1 = r\cos\theta. \tag{5}$$

The transformation from the system of coordinates ξ^1, η^1 to ξ, η has the form

$$\left.\begin{aligned} \xi_1 &= \xi_2^1 \cos\theta_0 \cos\varphi_0 - \xi_2^1 \sin\varphi_0 + \eta^1 \sin\theta_0 \cos\varphi_0, \\ \xi_2 &= \xi_1^1 \cos\theta_0 \sin\varphi_0 + \xi_2^1 \cos\varphi_0 + \eta^1 \sin\theta_0 \sin\varphi_0, \\ \eta &= -\xi_1^1 \sin\theta_0 + \eta^1 \cos\theta_0. \end{aligned}\right\} \tag{6}$$

Consider now a family of confocal ellipsoids $S(M, \tau)$ given in polar form by

$$r = \frac{\tau^2 - r^2(M, M_0)}{2[\tau - r(M, M_0)\cos\theta]}. \tag{7}$$

It is not difficult to write down parametric equations of the surface $S(M, \tau)$. For this purpose it is sufficient to substitute the last expression for r into (5) and the values obtained for ξ_1^1, ξ_2^1, η^1 into (6). This yields ξ_1, ξ_2, η as functions of the variables θ, φ and the parameter τ.

We now pass in formula (4) from the variables of integration ξ_1, ξ_2, η

to θ, φ, τ. Taking into account the parametric equations of the surface $S(M, \tau)$, we can easily establish the following formula:

$$\frac{dv_P}{r(P, M)\, r(P, M_0)} = \frac{2r^2(P, M_0)}{\tau^2 - r^2(M, M_0)} \sin\theta\, d\theta\, d\varphi\, d\tau.$$

By use of this we can write (4) in the form

$$\begin{aligned} v(M, t) = {} & \frac{1}{\pi[t^2 - r^2(M, M_0)]} \iint\limits_{S(M,t)} r^2(P, M_0)\, a(P)\, d\omega \\ & + \frac{1}{2\pi} \int\limits_{r(M, M_0)}^{t} \frac{d\tau}{\tau^2 - r^2(M, M_0)} \iint\limits_{S(M,\tau)} r^3(P, M_0)\, a(P) \\ & \times v(P, t - r(P, M))\, d\omega, \end{aligned} \tag{8}$$

where $d\omega = \sin\theta\, d\theta\, d\varphi$ is the element of solid angle with center at M_0.

We now transform the first integral in (8) assuming $a(P) \equiv a(\eta)$. By formula (7) the equation of the ellipsoid $S(M, t)$ has the form

$$r = \frac{t^2 - r^2(M, M_0)}{2[t - r(M, M_0) \cos\theta]}.$$

Introducing, for brevity, parameters ε and p defined by

$$\varepsilon = \frac{r(M, M_0)}{t}, \qquad p = \frac{t}{2}(1 - \varepsilon^2),$$

we can represent it in the form

$$r(\theta) = \frac{p}{1 - \varepsilon \cos\theta}. \tag{9}$$

The coordinate η of a point P lying on the ellipsoid $S(M, t)$ can be expressed in terms of its spherical coordinates according to (5) and (6) by the formula

$$\eta = r(\theta)\,(\cos\theta \cos\theta_0 - \sin\theta \sin\theta_0 \cos\varphi).$$

Passing in the first integral in (8) from the variable of integration φ to η, we find

$$\iint\limits_{S(M,t)} r^2(P, M_0)\, a(\eta) \sin\theta\, d\theta\, d\varphi$$

$$=2\int_0^\pi r^2(\theta)\sin\theta\,d\theta \int_{\eta_1(\theta)}^{\eta_2(\theta)} \frac{a(\eta)\,d\eta}{\sqrt{r^2(\theta)\sin^2\theta\sin^2\theta_0-[r(\theta)\cos\theta\cos\theta_0-\eta]^2}}$$

where

$$\left.\begin{aligned} n_1(\theta)&=r(\theta)\cos(\theta+\theta_0),\\ \eta_2(\theta)&=r(\theta)\cos(\theta-\theta_0). \end{aligned}\right\}$$

We now apply formula (9) from which follows

$$r^2(\theta)\sin\theta\,d\theta=-\frac{p}{\varepsilon}\,dr, \tag{10}$$

and we pass to the variable of integration r. Changing the order of integration in the iterated integral, we obtain

$$\iint\limits_{S(M,t)} r^2(P, M_0)\,a(\eta)\,d\omega$$

$$=2p\int_{\eta_1}^{\eta_2} a(\eta)\,d\eta\int_{r_1}^{r_2}$$

$$\cdot\frac{dr}{\sqrt{-r^2(1-\varepsilon^2\sin^2\theta_0)+2r(p+\varepsilon\eta\cos\theta_0)-[(p+\varepsilon\cos\theta_0)^2+\varepsilon^2\eta^2\sin^2\theta_0]}}.$$

In the last formula r_1 and r_2 are the roots of the quadratic trinomial in the denominator, namely,

$$r_k=\frac{p+\varepsilon\eta\cos\theta_0+(-1)^k\,\varepsilon\sin\theta_0\sqrt{(p+\varepsilon\eta\cos\theta_0)^2-\eta^2(1-\varepsilon^2\sin^2\theta_0)}}{1-\varepsilon^2\sin^2\theta_0}, \quad (k=1,2) \tag{11}$$

and η_1 and η_2 are those values of η for which $r_1=r_2$, i.e.

$$\eta_k=\frac{p}{1-\varepsilon^2}\left[\varepsilon\cos\theta_0+(-1)^k\sqrt{1-\varepsilon^2\sin^2\theta_0}\right], \quad (k=1,2). \tag{12}$$

From the above we find

$$\iint\limits_{S(M,t)} r^2(P, M_0)\,a(\eta)\,d\omega$$

$$= \frac{2p}{\sqrt{1-\varepsilon^2 \sin^2\theta_0}} \int\limits_{\eta_1}^{\eta_2} a(\eta)\, d\eta \int\limits_{r_1}^{r_2} \frac{dr}{\sqrt{(r_2-r)(r-r_1)}}$$

$$= \frac{2\pi p}{\sqrt{1-\varepsilon^2 \sin^2\theta_0}} \int\limits_{\eta_1}^{\eta_2} a(\eta)\, d\eta .$$

Reverting to Cartesian coordinates x and y, we can write (8) in the form

$$v(M,t) = \frac{1}{\sqrt{t^2-|x|^2}} \int\limits_{\frac{1}{2}(y-\sqrt{t^2-|x|^2})}^{\frac{1}{2}(y+\sqrt{t^2-|x|^2})} a(\eta)\, d\eta + \frac{1}{2\pi} \int\limits_{r(M,M_0)}^{t} \frac{d\tau}{\tau^2 - r^2(M,M_0)} \cdot \iint\limits_{S(M,\tau)} r^3(P,M_0)\, a(\eta)\, v(P, t-r(P,M))\, d\omega . \tag{13}$$

Differentiating this equality with respect to t, we find

$$\begin{aligned} v_t(M,t) = {} & \frac{t}{2(t^2-|x|^2)} \left[a\left(\frac{y+\sqrt{t^2-|x|^2}}{2}\right) + a\left(\frac{y-\sqrt{t^2-|x|^2}}{2}\right) \right] \\ & - \frac{t}{\sqrt{(t^2-|x|^2)^3}} \int\limits_{\frac{1}{2}(y-\sqrt{t^2-|x|^2})}^{\frac{1}{2}(y+\sqrt{t^2-|x|^2})} a(\eta)\, d\eta \\ & + \frac{1}{2\pi} \frac{1}{t^2 - r^2(M,M_0)} \cdot \iint\limits_{S(M,t)} r^3(P,M_0)\, a(\eta)\, v(P, t-r(P,M))\, d\omega \\ & + \frac{1}{2\pi} \int\limits_{r(M,M_0)}^{t} \frac{d\tau}{\tau^2 - r^2(M,M_0)} \cdot \iint\limits_{S(M,\tau)} r^3(P,M_0)\, a(\eta)\, v_t(P, t-r(P,M))\, d\omega . \end{aligned} \tag{14}$$

Introduce the function $w(M, t)$ defined by

$$w(M, t)=\frac{t^2-|x|^2}{t} v_t(M, t). \tag{15}$$

From (14) we obtain for the function $w(M, t)$ the equation

$$\begin{aligned} w(M, t)=&\tfrac{1}{2}\left[a\left(\frac{y+\sqrt{t^2-|x|^2}}{2}\right)+a\left(\frac{y-\sqrt{t^2-|x|^2}}{2}\right)\right] \\ &-\frac{1}{\sqrt{t^2-|x|^2}} \int\limits_{\frac{1}{2}(y-\sqrt{t^2-|x|^2})}^{\frac{1}{2}(y+\sqrt{t^2-|x|^2})} a(\eta)\, d\eta \\ &+\mathrm{B}(a(y)\, v(M, t))+\mathrm{C}(a(y)\, w(M, t)), \end{aligned} \tag{16}$$

where B and C are linear operators given by

$$\begin{aligned} \mathrm{B}v(M, t)=&\frac{t^2-|x|^2}{2\pi t\,[t^2-r^2(M, M_0)]} \\ &\cdot \iint\limits_{S(M,t)} r^3(P, M_0)\, v(P, t-r(P, M))\, d\omega, \end{aligned} \tag{17}$$

$$\begin{aligned} \mathrm{C}v(M, t)=&\frac{t^2-|x|^2}{2\pi t} \int\limits_{r(M,M_0)}^{t} \frac{d\tau}{\tau^2-r^2(M, M_0)} \\ &\cdot \iint\limits_{S(M,\tau)} \frac{r^3(P, M_0)\,[t-r(P, M)]}{[t-r(P, M)]^2-|\xi|^2} \\ &\cdot v(P, t-r(P, M))\, d\omega. \end{aligned} \tag{18}$$

To further abridge notation we introduce the linear operator A,

$$\begin{aligned} \mathrm{A}v(M, t)=&\frac{1}{2\pi} \int\limits_{r(M,M_0)}^{t} \frac{d\tau}{\tau^2-r^2(M, M_0)} \\ &\cdot \iint\limits_{S(M,\tau)} r^3(P, M_0)\, v(P, t-r(P, M))\, d\omega. \end{aligned} \tag{19}$$

Then (13) can be expressed as

$$v(M,t)=\frac{1}{\sqrt{t^2-|x|^2}}\int\limits_{\frac{1}{2}(y-\sqrt{t^2-|x|^2})}^{\frac{1}{2}(y+\sqrt{t^2-|x|^2})} a(\eta)\,d\eta+\mathrm{A}(a(y)\,v(M,t)). \qquad (13')$$

II. First of all we shall discuss the properties of the functions $v(M,t)$ and $w(M,t)$ determined by (13) and (16). Suppose that the function $a(y)$ is given and belongs to the class of continuous functions. Let us show that (13) and (16) then have a unique solution in the class of continuous functions which can be obtained by the method of successive approximations.[2]

Note that as follows from (13) and (16), there is a one-to-one correspondence between ellipsoid of revolution $S(M,t)$ and each point M and parameter t. Consider now the ellipsoid $S(M_1, T)$ where M_1 is an arbitrary point of the plane $y=0$ and T is an arbitrary number such that $T \geqslant r(M_0, M_1)$, and choose pairs M, t such that the ellipsoids corresponding to them do not go beyond the surface of the ellipsoid $S(M_1, T)$. To satisfy this condition it is necessary and sufficient that the set of M and t belong to the four-dimensional region

$$\mathscr{D}(M_1, T): \{r(M, M_0) \leqslant t \leqslant T - r(M, M_1)\}.$$

Denote by s the smaller semi-axis of the ellipsoid of revolution $S(M_1, T)$, i.e.

$$s=\tfrac{1}{2}\sqrt{T^2-r^2(M_0, M_1)}. \qquad (20)$$

Introduce also the notation

$$\left.\begin{aligned} \|v\| &= \max_{(M,t)\in\mathscr{D}(M_1,T)} |v(M,t)|, \\ \|w\| &= \max_{(M,t)\in\mathscr{D}(M_1,T)} |w(M,t)|, \\ \|a\| &= \max_{|y|\leqslant s} |a(y)|. \end{aligned}\right\} \qquad (21)$$

The following theorem holds.

Theorem 2.1. *To each given continuous function $a(y)$ there exists a unique continuous solution in the region $\mathscr{D}(M_1, T)$ for* (13) *and* (16). *More-*

[2] Equations (13) and (16) can be analyzed comparatively easily also by use of the resolvent for the telegraph equation constructed by B. M. Levitan [51].

over the solution satisfies the following inequalities:

$$\left.\begin{aligned}\|v\| &\leqslant \|a\|\; q(\|a\|\;\; s^2),\\ \|w\| &\leqslant \|a\|\; q_1(\|a\|\;\; s^2),\end{aligned}\right\} \tag{22}$$

where

$$\left.\begin{aligned} q(t) &= \sum_{k=0}^{\infty} \frac{t^k}{(k!)^2\,(k+1)},\\ q_1(t) &= 2+\tfrac{5}{4}(1+2\pi^2)\; t q(t).\end{aligned}\right\} \tag{23}$$

To prove the theorem we use the following formulas:

$$\left.\begin{aligned} &\mathrm{A}\left(\frac{t^2-r^2(M,M_0)}{4}\right)^k = \left(\frac{t^2-r^2(M,M_0)}{4}\right)^{k+1}\cdot\frac{1}{(k+1)\,(k+2)},\\ &\mathrm{B}\left(\frac{t^2-r^2(M,M_0)}{4}\right)^k = \begin{cases}\frac{1}{4}(t^2-|x|^2), & k=0,\\ \quad 0\quad, & k\neq 0\end{cases}\\ &\qquad k=0,1,2,\ldots\end{aligned}\right\} \tag{24}$$

$$\mathrm{C}(1) \leqslant \frac{\pi}{4}(t^2-|x|^2). \tag{25}$$

Therefore we shall pause to discuss their derivation. To derive the first two expressions, we use formula (10) which is

$$r^2(\theta)\sin\theta\, d\theta = -\frac{p}{\varepsilon}\, dr,$$

$$\left(\varepsilon = \frac{r(M,M_0)}{\tau},\quad p=\frac{\tau}{2}(1-\varepsilon^2)\right),$$

and the fact that on the surface $S(M,\tau)$, we have

$$r(P,M)=\tau-r(P,M_0).$$

Starting from this, we obtain

$$\frac{1}{2\pi}\iint\limits_{S(M,\tau)} r^3(P,M_0)\,[(t-r(P,M))^2-r^2(P,M_0)]^k\, d\omega = \frac{p}{\varepsilon}(t-\tau)^k$$

$$\cdot\int\limits_{p/(1+\varepsilon)}^{p/(1-\varepsilon)} r\;\,(t-\tau+2r)^k\, dr = \frac{\tau^2-r^2(M,M_0)}{8r(M,M_0)}\,(t-\tau)^k$$

$$\cdot\left[\frac{(t+r(M,M_0))^{k+2}-(t-r(M,M_0))^{k+2}}{k+2}\right.$$

$$-(t-\tau)\frac{(t+r(M,M_0))^{k+1}-(t-r(M,M_0))^{k+1}}{k+1}\Bigg]. \quad k=0,1,2,\dots.$$

Using the above equality and formulas (17) and (19), we find

$$\mathrm{A}\left(\frac{t^2-r^2(M,M_0)}{4}\right)^k$$

$$=\frac{1}{2\pi}\int\limits_{r(M,M_0)}^{t}\frac{d\tau}{\tau^2-r^2(M,M_0)}$$

$$\cdot\iint\limits_{S(M,\tau)} r^3(P,M_0)\,[(t-r(P,M))^2-r^2(P,M_0)]^k\,d\omega$$

$$=\frac{1}{8r(M,M_0)}\left[\frac{(t+r(M,M_0))^{k+2}-(t-r(M,M_0))^{k+2}}{k+2}\int\limits_{r(M,M_0)}^{t}(t-\tau)^k d\tau\right.$$

$$\left.-\frac{(t+r(M,M_0))^{k+1}-(t-r(M,M_0))^{k+1}}{k+1}\int\limits_{r(M,M_0)}^{t}(t-\tau)^{k+1}\,d\tau\right]$$

$$=\left(\frac{t^2-r^2(M,M_0)}{4}\right)^{k+1}\frac{1}{(k+1)(k+2)}, \quad k=0,1,2,\dots$$

$$\mathrm{B}\left(\frac{t^2-r^2(M,M_0)}{4}\right)^k=\frac{t^2-|x|^2}{2\pi t\,[t^2-r^2(M,M_0)]}$$

$$\cdot\iint\limits_{S(M,t)} r^3(P,M_0)\,[(t-r(P,M))^2-r^2(P,M_0)]^k\,d\omega$$

$$=\frac{t^2-|x|^2}{8r(M,M_0)\,t}\left[(t-\tau)^k\frac{(t+r(M,M_0))^{k+2}-(t-r(M,M_0))^{k+2}}{k+2}\right.$$

$$\left.-(t-\tau)^{k+1}\frac{(t+r(M,M_0))^{k+1}-(t-r(M,M_0))^{k+1}}{k+1}\right]_{\tau=t}$$

$$=\begin{cases}\frac{1}{4}(t^2-|x|^2, & k=0\\ \quad 0\quad, & k=1,2,3,\dots.\end{cases}$$

Thus formulas (24) are valid.

We now prove inequality (25). To this end, we first reduce C(1), the left-hand side of the inequality, to the form

$$C(1)=\frac{t^2-|x|^2}{2\pi t}\int_{r(M,M_0)}^{t}\frac{d\tau}{\sqrt{\tau^2-|x|^2}}\int_{\eta_1}^{\eta_2}d\eta\int_{r_1}^{r_2}\frac{r(t-\tau+r)}{(t-\tau)(t-\tau+2r)+\eta^2}\cdot\frac{dr}{\sqrt{(r_2-r)(r-r_1)}},$$

where η_k and $r_k (k=1, 2)$ are given by formulas (11) and (12). It is easy to convince ourselves that the expression

$$\frac{r(t-\tau+r)}{(t-\tau)(t-\tau+2r)+\eta^2}$$

is a monotone increasing function of $r(r\geqslant 0)$. At the same time $r\leqslant\tau$. Therefore

$$\frac{r(t-\tau+r)}{(t-\tau)(t-\tau+2r)+\eta^2}\leqslant\frac{\tau t}{t^2-\tau^2+\eta^2}.$$

From this we find

$$\int_{\eta_1}^{\eta_2}d\eta\int_{r_1}^{r_2}\frac{r(t-\tau+r)}{(t-\tau)(t-\tau+2r)+\eta^2}\cdot\frac{dr}{\sqrt{(r_2-r)(r-r_1)}}$$
$$\leqslant\tau t\int_{\eta_1}^{\eta_2}\frac{d\eta}{t^2-\tau^2+\eta^2}\int_{r_1}^{r_2}\frac{dr}{\sqrt{(r_2-r)(r-r_1)}}$$
$$=\frac{\pi\tau t}{\sqrt{t^2-\tau^2}}\left[\operatorname{arc\,tan}\frac{\eta_2}{\sqrt{t^2-\tau^2}}-\operatorname{arc\,tan}\frac{\eta_1}{\sqrt{t^2-\tau^2}}\right]\leqslant\frac{\pi^2\tau t}{\sqrt{t^2-\tau^2}}.$$

Using this inequality we obtain

$$C(1)\leqslant\pi\frac{t^2-|x|^2}{2}\int_{r(M,M_0)}^{t}\frac{\tau\,d\tau}{\sqrt{(\tau^2-|x|^2)(t^2-\tau^2)}}$$
$$\leqslant\pi\frac{t^2-|x|^2}{2}\int_{r(M,M_0)}^{t}\frac{\tau\,d\tau}{\sqrt{[\tau^2-r^2(M,M_0)][t^2-\tau^2]}}=\pi^2\frac{t^2-|x|^2}{4}.$$

We further note that from (18) and (19) results the following interconnection between the operators A and C:

$$\mathrm{C}\left(\frac{t^2-|x|^2}{t}\,v(M,t)\right)=\frac{t^2-|x|^2}{t}\,\mathrm{A}v(M,t). \tag{26}$$

We proceed now to prove the theorem formulated above. Consider successive approximations for (13) defined by the scheme

$$\begin{gathered}
v_0(M,t)=\frac{1}{\sqrt{t^2-|x|^2}}\int\limits_{\frac{1}{2}(y-\sqrt{t^2-|x|^2})}^{\frac{1}{2}(y+\sqrt{t^2-|x|^2})} a(\eta)\,d\eta,\\
v_n(M,t)=v_0(M,t)+\mathrm{A}(a(y)\,v_{n-1}(M,t))\\
n=1,2,3,\ldots.
\end{gathered} \tag{27}$$

It is obvious that each function $v_n(M, t)$ is continuous in the region $\mathscr{D}(M_1, T)$. Simultaneously equality (24) and the linearity of the operator A imply the following estimates in this region:

$$\begin{aligned}
&|v_0(M,t)|\leqslant\|a\|,\\
&|v_1(M,t)|\leqslant|v_0(M,t)|+\mathrm{A}(\|a\|\;|v_0(M,t)|)\\
&\leqslant\|a\|(1+\mathrm{A}\,(\|a\|))=\|a\|\left[1+\frac{\|a\|}{(1!)^2\cdot 2}\,\frac{t^2-r^2(M,M_0)}{4}\right],\\
&|v_2(M,t)|\leqslant|v_0(M,t)|+\mathrm{A}\,(\|a\|\;|v_1(M,t)|)\\
&\leqslant\|a\|\left\{1+\mathrm{A}\left(\|a\|\left[1+\frac{\|a\|}{(1!)^2\cdot 2}\,\frac{t^2-r^2(M,M_0)}{4}\right]\right)\right\}\\
&=\|a\|\left[1+\frac{\|a\|}{(1!)^2\cdot 2}\,\frac{t^2-r^2(M,M_0)}{4}+\frac{\|a\|^2}{(2!)^2\cdot 3}\left(\frac{t^2-r^2(M,M_0)}{4}\right)^2\right],\\
&\cdots\cdots\cdots\cdots\cdots\cdots\\
&|v_n(M,t)|\leqslant\|a\|\sum_{k=0}^{n}\frac{\|a\|^k}{(k!)^2\,(k+1)}\left(\frac{t^2-r^2(M,M_0)}{4}\right)^k.
\end{aligned} \tag{28}$$

If we write the Neumann series $v_0(M,t)+\sum_{n=0}^{\infty}[v_n(M,t)-v_{n-1}(M,t)]$ for (13) corresponding to scheme (27), its partial sum will coincide with the function $v_n(M, t)$ because of the linearity of (13); hence, this series is majorized by

$$\|a\|\sum_{k=0}^{\infty}\frac{\|a\|^k}{(k!)^2\,(k+1)}\left(\frac{t^2-r^2(M,M_0)}{4}\right)^k.$$

This, in turn, is majorized for all $(M, t)\in\mathscr{D}(M_1, T)$ by the convergent series

$$\|a\| \sum_{k=0}^{\infty} \frac{\|a\|^k}{(k!)^2 (k+1)} s^{2k} = \|a\| \; q(\|a\| \; s^2).$$

This last statement follows from the fact that in the region $\mathscr{D}(M_1, T)$,

$$t^2 - r^2(M, M_0) \leqslant T^2 - r^2(M, M_1) = 4s^2.$$

It is not difficult to show that the sequence $v_n(M, t)$, $n=0, 1, 2, \ldots$, converges by itself. Indeed, using formulas (27) and making estimates similar to the above we find

$$|v_n(M, t) - v_{n-1}(M, t)| \leqslant \frac{\|a\|^{n+1}}{(n!)^2 (n+1)} \left(\frac{t^2 - r^2(M, M_0)}{4}\right)^n \leqslant \frac{\|a\|^{n+1} s^{2n}}{(n!)(n+1)}.$$

This implies the validity of our statement. Thus the Neumann series converges uniformly and hence, its sum is a continuous function in the region $\mathscr{D}(M_1, T)$. As usual, it is easy to prove that the sum of the series is the solution of (13) and that the solution is unique. Passing to the limit in the last inequality in (28) we obtain an estimate for the solution, namely,

$$|v(M, t)| \leqslant \|a\| \sum_{k=0}^{\infty} \frac{\|a\|^k}{(k!)^2 (k+1)} \left(\frac{t^2 - r^2(M, M_0)}{4}\right)^k.$$

The first formula of (22) then becomes obvious.

We now consider (16). The function $v(M, t)$ can be regarded as known. For (16) we construct successive approximations similar to scheme (27):

$$\left.\begin{aligned} w_0(M, t) &= \tfrac{1}{2}\left[a\left(\frac{y+\sqrt{t^2-|x|^2}}{2}\right) + a\left(\frac{y-\sqrt{t^2-|x|^2}}{2}\right)\right] \\ &\quad - \frac{1}{\sqrt{t^2-|x|^2}} \int_{\frac{1}{2}(y-\sqrt{t^2-|x|^2})}^{\frac{1}{2}(y+\sqrt{t^2-|x|^2})} a(\eta)\, d\eta + \mathrm{B}(a(y)\, v(M, t)), \\ w_n(M, t) &= w_0(M, t) + \mathrm{C}(a(y)\, w_{n-1}(M, t)), \qquad n=1, 2, 3, \ldots. \end{aligned}\right\} \quad (29)$$

It is easy to prove continuity in the region $\mathscr{D}(M_1, T)$ of each function $w_n(M, t)$ on the basis of inequality (25). Let us estimate in this region the absolute value of the terms of the sequence $w_n(M, t)$ $(n=0, 1, 2, \ldots)$. Using formulas (24), (25) and (26) we find

$$\begin{aligned} |w_0(M, t)| &\leqslant 2\,\|a\| + \mathrm{B}(\|a\| \cdot |v(M, t)|) \\ &\leqslant \|a\| \left[2 + \mathrm{B}\left(\|a\| \sum_{k=0}^{\infty} \frac{\|a\|^k}{(k!)^2 (k+1)} \left(\frac{t^2 - r^2(M, M_0)}{4}\right)^k\right)\right] \end{aligned}$$

$$= \|a\| \left[2 + \sum_{k=0}^{\infty} \frac{\|a\|^{k+1}}{(k!)^2 (k+1)} \mathrm{B}\left(\frac{t^2 - r^2(M, M_0)}{4}\right)^k\right]$$

$$= \|a\| \left[2 + \|a\| \frac{t^2 - |x|^2}{4}\right]$$

$$|w_1(M, t)| \leqslant |w_0(M, t)| + \mathrm{C}(\|a\| \, w_0(M, t)|)$$

$$\leqslant \|a\| \left\{2 + \|a\| \frac{t^2 - |x|^2}{4} + \mathrm{C}\left(\|a\| \cdot \left[2 + \|a\| \frac{t^2 - |x|^2}{4}\right]\right)\right\}$$

$$= \|a\| \left\{2[1 + \|a\| \, \mathrm{C}\,(1)] + \|a\| \left[\frac{t^2 - |x|^2}{4} + \mathrm{C}\left(\frac{t^2 - |x|^2}{4}\right)\right]\right\}$$

$$\leqslant \|a\| \left\{2\left[1 + \pi^2 \|a\| \frac{t^2 - |x|^2}{4}\right] + \|a\| \frac{t^2 - |x|^2}{4}\left[1 + \frac{\|a\|}{(1!)^2 \cdot 2}\right.\right.$$

$$\left.\left.\cdot \frac{t^2 - r^2(M, M_0)}{4}\right]\right\}$$

$$|w_2(M, t)| \leqslant |w_0(M, t)| + \mathrm{C}\,(\|a\| \cdot |w_1(M, t)|)$$

$$\leqslant \|a\| \left\{2 + \|a\| \frac{t^2 - |x|^2}{4} + \mathrm{C}\left(\|a\| \left[2\left(1 + \pi^2 \|a\| \frac{t^2 - |x|^2}{4}\right)\right.\right.\right.$$

$$\left.\left.\left. + \|a\| \frac{t^2 - |x|^2}{4}\left(1 + \frac{\|a\|}{(1!)^2 \cdot 2} \frac{t^2 - r^2(M, M_0)}{4}\right)\right]\right)\right\}$$

$$= \|a\| \left\{2\left[1 + \|a\| \, \mathrm{C}\left(1 + \pi^2 \|a\| \frac{t^2 - |x|^2}{4}\right)\right] + \|a\| \left[\frac{t^2 - |x|^2}{4}\right.\right.$$

$$\left.\left. + \mathrm{C}\left(\frac{t^2 - |x|^2}{4} \sum_{k=0}^{1} \frac{\|a\|^k}{(k!)^2 (k+1)} \left(\frac{t^2 - r^2(M, M_0)}{4}\right)^k\right)\right]\right\}$$

$$\leqslant \|a\| \left\{2 + \|a\| \frac{t^2 - |x|^2}{4} \left[2\pi^2 \sum_{k=0}^{1} \frac{\|a\|^k}{(k!)^2 (k+1)} \left(\frac{t^2 - r^2(M, M_0)}{4}\right)^k\right.\right.$$

$$\left.\left. + \sum_{k=0}^{2} \frac{\|a\|^k}{(k!)^2 (k+1)} \left(\frac{t^2 - r^2(M, M_0)}{4}\right)^k\right]\right\},$$

. . . .

$$|w_n(M, t)| \leqslant \|a\| \left\{2 + \|a\| \frac{t^2 - |x|^2}{4} \left[2\pi^2 \sum_{k=0}^{n-1} \frac{\|a\|^k}{(k!)^2 (k+1)} \left(\frac{t^2 - r^2(M, M_0)}{4}\right)^k\right.\right.$$

$$\left.\left. + \sum_{k=0}^{n} \frac{\|a\|^k}{(k!)^2 (k+1)} \left(\frac{t^2 - r^2(M, M_0)}{4}\right)^k\right]\right\}.$$

Using the above estimates, we can easily prove in a similar way the convergence of the successive approximations to a unique continuous solution of (16) in $\mathscr{D}(M_1, T)$. Passing to the limit in the last inequality and taking into account that the inequalities

$$t^2 - r^2(M, M_0) \leqslant 4s^2, \ |\eta| \leqslant s,$$

hold in the region $\mathscr{D}(M_1, T)$, we obtain the second formula in (22).

Note that the function $q(t)$ determined by the first formula in (23) is related to the derivative of the Bessel function of zero order $\mathrm{J}_0(z)$. Indeed, $\mathrm{J}_0(z)$ is given by the series

$$\sum_{k=0}^{\infty} \frac{(-1)^k}{(k!)^2} \left(\frac{z}{2}\right)^{2k} = \mathrm{J}_0(z).$$

Simultaneously the expression for $q(t)$ can easily be transformed into

$$q(t) = \frac{d}{dt} \sum_{k=0}^{\infty} \frac{t^{k+1}}{[(k+1)!]^2} = \frac{d}{dt} \sum_{k=0}^{\infty} \frac{t^k}{(k!)^2}.$$

Comparing the two expressions we find

$$q(t) = \frac{d}{dt} \mathrm{J}_0(2i\sqrt{t}).$$

III. We now consider the inverse problem. Assume as before that $a(y)$ is a continuous function. Suppose at the point $M_1(x_1^1, x_2^1, 0)$ the generalized solution of (1) under conditions (2) is known and, hence, the function $v(M_1, t)$ is known. Find the function $a(y)$. Let

$$v(M_1, t) + w(M_1, t) = f(t).$$

From the theorem proved above it follows that the function $f(t)$ is continuous on the half-line $t \geqslant r(M_1, M_0)$. We add the right hand sides and the left hand sides of (13) and (16) and we let M tend to M_1 in the equality obtained and $t \to T$. Making use of the evenness of $a(y)$ we obtain the equation

$$a(s) = f(T) - [\mathrm{A}(a(y)\, v(M, t)) + \mathrm{B}(a(y)\, v(M, t)) + \mathrm{C}(a(y)\, w(M, t))]_{M_1, T}, \tag{30}$$

where s and T are related by formula (20) and the symbol $[\ ]_{M_1, T}$ means that the function in brackets is evaluated at the point M_1, T.

Introduce the operator U defined by

$$\mathrm{U}a(y) \equiv f(T) - [\mathrm{A}(a(y)\, v(M, t)) + \mathrm{B}(a(y)\, v(M, t)) + \mathrm{C}(a(y)\, w(M, t))]_{M_1, T}, \tag{31}$$

in which the functions $v(M, t)$ and $w(M, t)$ on the right-hand side of the equation are regarded as composite functions of the variables $M, t, a(y)$ determined by (13) and (16). Eq. (30) can be written in the more suitable form

$$a(s)=\mathrm{U}a(y). \tag{32}$$

Consider the region bounded by the curvilinear trapezoid

$$0\leqslant s\leqslant s_1,\ |a(s)-f(T)|\leqslant R,$$

where R is an arbitrary positive number. Let

$$\|f\|=\max_{r(M,M_0)\leqslant t\leqslant T_1}|f(t)|;\qquad T_1=\sqrt{4s_1^2+r^2(M_1,M_0)}.$$

The values $\|v\|$, $\|w\|$ and $\|a\|$ introduced before by formula (21) depend monotonically on s as a parameter in the interval $[0, s_1]$. Let

$$0<s^*<\min(s_1, s_2, s_3),$$

where

$$s_2=\sqrt{\frac{t_2}{R+\|f\|}},\qquad s_3=\sqrt{\frac{t_3}{R+\|f\|}}, \tag{33}$$

t_2 being a root of the equation

$$t\left[\tfrac{3}{2}q(t)+\pi^2 q_1(t)\right]=\frac{R}{R+\|f\|}, \tag{34}$$

and t_3 a root of the equation

$$\begin{aligned}t\{q(t)\,&[3+\tfrac{1}{2}tq(t)]+\pi^2 q_1(t)+\pi^2(1+\tfrac{5}{4}\pi^2 tq(t))\\&\cdot[2+tq(t)\,(2+\tfrac{1}{2}q(t))+\pi^2 tq_1(t)]\}=1.\end{aligned} \tag{35}$$

Theorem 2.2. *In the region*

$$\mathscr{D}:\{0\leqslant s\leqslant s^*,\ |a(s)-f(T)|\leqslant R\}$$

equation (32) *has a unique continuous solution.*

Consider the set of the functions $\mathfrak{M}$ in the space C belonging to the region $\mathscr{D}$. We prove that U is a contracting operator on the set $\mathfrak{M}$. For this purpose we shall show first that the operator U transforms $\mathfrak{M}$ into itself, i.e. $a(s)\in\mathfrak{M}$ implies $\mathrm{U}a(y)\in\mathfrak{M}$. Indeed, for any continuous function $a(y)$, (13) and (16) determine, by Theorem 2.1, continuous functions $v(M, t)$ and $w(M, t)$. The function $\mathrm{U}a(y)$ given by (31) will also be continuous in this case. Moreover, using formulas (22), (24) and (25) we find

$$\begin{aligned}|\mathrm{U}a(y)-f(T)|&\leqslant\|a\|\,\{\|v\|\,[\mathrm{A}(1)+\mathrm{B}(1)]+\|w\|\,\mathrm{C}(1)\}\\&\leqslant s^2\,\|a\|^2\,\left[\tfrac{3}{2}q(\|a\|\,s^2)+\pi^2 q_1(\|a\|\,s^2)\right].\end{aligned}$$

Note that the functions $q(t)$ and $q_1(t)$ increase monotonically with increasing t, and the fact that $a(s)\in\mathfrak{M}$ implies

$$\|a\|\leqslant R+\|f\|. \tag{36}$$

Therefore for all $s\in[0, s^*]$ the inequality will only be sharpened if we substitute s_2 for s and replace $\|a\|$ by the expression $R+\|f\|$. Carrying this out and using formulas (33) and (34) we obtain

$$|\mathrm{U}a(y)-f(T)|\leqslant(R+\|f\|)\,[t(\tfrac{3}{2}q(t)+\pi^2 q_1(t))]_{t=s_2^2(R+\|f\|)}=R.$$

Hence we have proved that $\mathrm{U}a(s)\in\mathfrak{M}$.

Consider now any two functions $a(s)$ and $\tilde{a}(s)$ belonging to $\mathfrak{M}$. Let us estimate in space C the distance between their images $\mathrm{U}a(y)$ and $\mathrm{U}\tilde{a}(y)$. The solution of (13) and (16) corresponding to the function $\tilde{a}(y)$ will be denoted by $\tilde{v}(M, t)$ and $\tilde{w}(M, t)$. From (31) we find

$$\begin{aligned}|\mathrm{U}a(y)-\mathrm{U}\tilde{a}(y)|&\leqslant \mathrm{A}(\|av-\tilde{a}\tilde{v}\|)+\mathrm{B}(\|av-\tilde{a}\tilde{v}\|)+\mathrm{C}(\|av-\tilde{a}\tilde{v}\|)\\&\leqslant s^2\,[\tfrac{3}{2}\,\|av-\tilde{a}\tilde{v}\|+\pi^2\,\|aw-\tilde{a}\tilde{w}\|]\\&\leqslant s^2\,[\|a-\tilde{a}\|\,(\tfrac{3}{2}\,\|v\|+\pi^2\,\|w\|)\\&\quad+\|\tilde{a}\|\,(\tfrac{3}{2}\,\|v-\tilde{v}\|+\pi^2\,\|w-\tilde{w}\|)]\\&\leqslant s^2[\|a-\tilde{a}\|\cdot\|a\|\,[\tfrac{3}{2}q(\|a\|\,s^2)+\pi^2 q_1(\|a\|\,s^2)]\\&\quad+\|\tilde{a}\|\,(\tfrac{3}{2}\,\|v-\tilde{v}\|+\pi^2\,\|w-\tilde{w}\|)].\end{aligned}\tag{37}$$

Using (13) and (16), we introduce inequalities for $\|v-\tilde{v}\|$ and $\|w-\tilde{w}\|$ in terms of $\|a-\tilde{a}\|$. Considering (13) we obtain

$$\begin{aligned}|v(M, t)-\tilde{v}(M, t)|&\leqslant\frac{1}{\sqrt{t^2-|x|^2}}\int\limits_{\frac{1}{2}(y-\sqrt{t^2-|x|^2})}^{\frac{1}{2}(y+\sqrt{t^2-|x|^2})}|a(\eta)-\tilde{a}(\eta)|\,d\eta\\&\quad+\mathrm{A}(|a(y)\,v(M, t)-\tilde{a}(y)\,\tilde{v}(M, t)|)\leqslant\|a-\tilde{a}\|\\&\quad+\mathrm{A}(\|a-\tilde{a}\|\,\|\tilde{v}\|+\|\tilde{a}\|\,|v(M, t)-\tilde{v}(M, t)|).\end{aligned}$$

Substituting estimate (22) for $\|v\|$ into the last inequality we obtain a linear inequality for the difference $|v(M, t)-\tilde{v}(M, t)|$,

$$\begin{aligned}&|v(M, t)-\tilde{v}(M, t)|\\&\leqslant\|a-\tilde{a}\|\left[1+\frac{s^2}{2}\,\|a\|\,q(\|a\|\,s^2)\right]+\|\tilde{a}\|\,\mathrm{A}(|v(M, t)-\tilde{v}(M, t)|).\end{aligned}$$

Applying the method of successive approximations to it, we obtain

$$\|v-\tilde{v}\|\leqslant\|a-\tilde{a}\|\left[1+\frac{s^2}{2}\,\|a\|\,q(\|a\|\,s^2)\right]q(\|\tilde{a}\|\,s^2). \tag{38}$$

We now estimate $\|w-\tilde{w}\|$. From (16) we obtain analogously to the foregoing, the inequality

$$|w(M,t)-\tilde{w}(M,t)| \leqslant \|a-\tilde{a}\| \, [2+\pi^2 s^2 q_1(\|a\| \; s^2)] \\ + s^2 \|av-\tilde{a}\tilde{v}\| + \|\tilde{a}\| \, \mathrm{C}(|w(M,t)-\tilde{w}(M,t)|).$$

The middle term on the right-hand side of the inequality can be estimated on the basis of inequality (28), as follows:

$$\|av-\tilde{a}\tilde{v}\| \leqslant \|a-\tilde{a}\| \, \|v\| + \|\tilde{a}\| \, \|v-\tilde{v}\| \\ \leqslant \|a-\tilde{a}\| \left\{ \|a\| \, q(\|a\| \; s^2) + \|\tilde{a}\| \right. \\ \left. \cdot \left[1+\frac{s^2}{2} \|a\| \, q(\|a\| \; s^2)\right] q(\|\tilde{a}\| \; s^2) \right\}.$$

Finally for the difference $|w(M,t)-\tilde{w}(M,t)|$ we obtain

$$|w(M,t)-\tilde{w}(M,t)| \\ \leqslant \|a-\tilde{a}\| \, \varphi(\|a\| \, s^2, \, \|\tilde{a}\| \, s^2) \; + \|\tilde{a}\| \, \mathrm{C}(|w(M,t)-\tilde{w}(M,t)|), \tag{39}$$

where

$$\varphi(\xi,\eta) = 2+\xi[q(\xi)+\pi^2 q_1(\xi)] + \eta q(\eta)\,(1+\tfrac{1}{2}\xi q(\xi)). \tag{40}$$

Solving inequality (39) by the method of successive approximations we obtain

$$\|w-\tilde{w}\| \leqslant \|a-\tilde{a}\| \, \varphi(\|a\| \, s^2, \, \|\tilde{a}\| \, s^2) \, [1+\tfrac{5}{4}\pi^2 \|\tilde{a}\| \, s^2 q(\|\tilde{a}\| \; s^2)] \tag{41}$$

Substituting the estimates (38) and (41) into the inequality, we obtain

$$|\mathrm{U}a(y)-\mathrm{U}\tilde{a}(y)| \\ \leqslant \|a-\tilde{a}\| \, s^2 \left\{ \|a\| \, [\tfrac{3}{2} q(\|a\| \cdot s^2) + \pi^2 q_1(\|a\| \; s^2)] \right. \\ + \|\tilde{a}\| \left[\tfrac{3}{2} q(\|\tilde{a}\| \; s^2) \left(1+\frac{s^2}{2} \|a\| \, q(\|a\| \; s^2)\right) \right. \\ + \pi^2 \varphi(\|a\| \, s^2, \, \|\tilde{a}\| \, s^2) \\ \left. \left. \cdot (1+\tfrac{5}{4}\pi^2 \|\tilde{a}\| \, s^2 q(\|\tilde{a}\| \; s^2)) \right] \right\}. \tag{42}$$

The functions $a(s)$ and $a(s)$ belong to the set $\mathfrak{M}$ and hence, inequality (46) holds for each of them. Note that the function in braces in (42) increases monotonically with increasing s. Therefore, if $\|a\|$ and $\|\tilde{a}\|$

are replaced in this equality by $R+\|f\|$ and s by s^*, the inequality is only sharpened. Doing this and using expression (40) for $\varphi(\xi, \eta)$ we find

$$|\mathrm{U}a(y)-\mathrm{U}\tilde{a}(y)| \leqslant \|a-\tilde{a}\|\, \psi[(s^*)^2(R+\|f\|)], \tag{43}$$

where

$$\psi(t)=t\{q(t)[3+\tfrac{1}{2}tq(t)]+\pi^2 q_1(t) + \pi^2(1+\tfrac{5}{4}\pi^2 tq(t))[2+tq(t)(2+\tfrac{1}{2}tq(t))+\pi^2 tq_1(t)]\}.$$

The function $\psi(t)$ is non-negative on the ray $t\geqslant 0$ and increases monotonically with increasing t. Because of this, for $t^*<t_3$,

$$\frac{\psi(t^*)}{\psi(t_3)}<1, \qquad t^*=(s^*)^2(R+\|f\|).$$

Recalling that t_3 is the solution of (35), i.e.

$$\psi(t)=1,$$

we obtain

$$\psi[(s^*)^2(R+\|f\|)]<1. \tag{44}$$

From inequalities (43) and (44) it follows that in the function space C the distance between the functions $\mathrm{U}a(y)$ and $\mathrm{U}\tilde{a}(y)$ is less than that between $a(s)$ and $\tilde{a}(s)$. Hence, the operator U is a contracting operator on the set $\mathfrak{M}$. But then by Banach's theorem, the operator U has only one fixed point on the set $\mathfrak{M}$, i.e. there exists a unique solution to (32), as asserted.

It should be noted that by Banach's theorem the solution of this equation can be obtained by the method of successive approximations, where any function belonging to $\mathfrak{M}$, in particular $f(T)$, can be taken as the zero approximation.

Remark. The number s^* appearing in the statement of the theorem, is dependent for given $f(t)$ only on R. Generally speaking, this dependence is of the following nature: if R is small, s^* is small too; with the increase of R, s^* also increases up to a maximum value s^{**} and then decreases again. The number s^{**} characterizes the maximum interval on which Theorem 2.2 guarantees existence and uniqueness in the problem of determining the function $a(y)$ from $f(t)$.

Note also that $v(M_1, t)\equiv 0$ implies $f(t)\equiv 0$. Moreover one can see from (33)–(35) that in this case, if $R\to 0$ and $s_1\to\infty$ then s^* can be taken to be an arbitrarily large number. Thus in this case $a(y)\equiv 0 (0\leqslant y<\infty)$.

This implies in particular, that if the function $v(M, t)$ is equal to zero for all t at least at one point of the plane $y=0$, then it is identically zero.

Theorem 2.2 has the same local character as the existence and uniqueness theorem for ordinary differential equations. Just as for ordinary differential equations, having constructed a solution on the interval $[0, s^*]$, one can extend it to $[s^*, s_1^*]$. For this purpose, (31) must be recalculated using the fact that in the region $\mathscr{D}(M_1, T^*)$ $(T^*=\sqrt{4(s^*)^2+r^2(M_1, M_0)})$ the functions $a(y)$, $v(M, t)$, $w(M, t)$ are already known. We again get an equation of type (32) with a small parameter. We shall not consider here how far the solution can be extended in this manner.

3. Linearized Inverse Problem for the Telegraph Equation

In the previous section we considered the problem of determining the coefficient $a(M)$ in the telegraph equation when $a(M)\equiv a(y)$. Now, $a(M)$ will be considered to be simply a continuous function of the point M and the solution of the inverse problem will be given in a linearized setting.

The essense of the linearization method is as follows. Suppose that the function $a(M)$ can be represented in the form

$$a(M)=a_0(M)+a_1(M) \tag{1}$$

where $a_0(M)$ is a known function and belongs to space C, and the function $a_1(M)$ is small in the norm in C. The problem is to find $a_1(M)$. We make use of the smallness of $a_1(M)$ to linearize (2.8) which is equivalent to (2.1) under condition (2.2). To make things clearer we introduce a parameter λ by

$$a(M)=a_0(M)+\lambda a_1(M). \tag{2}$$

For $\lambda=1$ formula (2) coincides with (1). We can represent the solution of (2.8) as a power series in λ,

$$v(M, M_0, t)=v_0(M, M_0, t)+\lambda v_1(M, M_0, t)+\lambda^2 v_2(M, M_0, t)+\cdots \tag{3}$$

Replacing the functions $a(M)$ and $v(M, M_0, t)$ in (2.8) by their expressions from (2) and (3) and collecting terms with the same powers of λ we find equations for the terms of series (3). They are

$$v_0(M, M_0, t)=\frac{1}{\pi[t^2-r^2(M, M_0)]}\iint\limits_{S(M, M_0, t)} r^2(P, M_0)\, a_0(P)\, d\omega$$
$$+\mathrm{A}(a_0(M)\, v_0(M, M_0, t)),$$

$$v_1(M, M_0, t)=\frac{1}{\pi[t^2-r^2(M, M_0)]}\iint\limits_{S(M, M_0, t)} r^2(P, M_0)\, a_1(P)\, d\omega \tag{4}$$

$$+\mathrm{A}(a_1(M)\, v_0(M, M_0, t)+a_0(M)\, v_1(M, M_0, t)),$$

. . . .

$$v_n(M, M_0, t)=\mathrm{A}(a_1(M)\, v_{n-1}(M, M_0, t)+a_0(M)\, v_n(M, M_0, t)),$$
$$(n=2, 3, \ldots)$$

where A is the operator given by (2.19). The meaning of the parameter M_0 is clear from (2.1).

Let $a_0(M)\equiv 0$ and hence, $a_1(M)\equiv a(M)$. Formulas (4) become slightly simplified. In particular, $v_0(M, M_0, t)\equiv 0$. The expressions for $v_n(M, M_0, t)$ become

$$v_1(M, M_0, t)=\frac{1}{\pi[t^2-r^2(M, M_0)]}\iint\limits_{S(M, M_0, t)} r^2(P, M_0)\, a(P)\, d\omega, \tag{5}$$

. . . .

$$v_n(M, M_0, t)=\mathrm{A}(a(M)\, v_{n-1}(M, M_0, t)),\ (n=2, 3, \ldots).$$

Let us estimate the difference between the exact solution of the problem $v(M, M_0, t)$ and its linear approximation $v_1(M, M_0, t)$. For this purpose consider a sphere of radius R in our space and denote by $\mathscr{D}(M_0, R)$ the point set M, t of the four-dimensional region,

$$r(M, M_0)\leqslant t\leqslant 2R-r(M, M_0).$$

Introduce the following notation similar to (2.21):

$$\left.\begin{aligned} \|v\| &= \max_{M,\, t\in\mathscr{D}(M_0, R)} |v(M, M_0, t)|, \\ \|a\| &= \max_{r(M, M_0)\leqslant R} |a(M)|. \end{aligned}\right\} \tag{6}$$

Then, taking into account that the projection of the set $\mathscr{D}(M_0, R)$ into x_1, x_2, y space is a sphere of radius R with center at M_0, we find

$$\|v_1\|\leqslant \|a\|\, \frac{1}{\pi[t^2-r^2(M, M_0)]}\iint\limits_{S(M, M_0, t)} r^2(P, M_0)\, d\omega=\|a\|. \tag{7}$$

It is not difficult to estimate the other terms of series (3). Using formula (2.24) and inequality (7) we find easily

$$|v_2(M, M_0, t)|\leqslant \|a\|\ \mathrm{A}(|v_1(M, M_0, t)|)\leqslant \|a\|^2\ \frac{t^2-r^2(M, M_0)}{4}\ \frac{1}{(1!)^2\cdot 2},$$

$$|v_3(M, M_0, t)|\leqslant \|a\|\ \mathrm{A}(|v_2(M, M_0, t)|)\leqslant \|a\|^3 \left(\frac{t^2-r^2(M, M_0)}{4}\right)^2 \frac{1}{(2!)^2\, 3},$$

. . . .

$$|v_n(M, M_0, t)| \leqslant \|a\| \, \mathrm{A}(|v_{n-1}(M, M_0, t)|) \leqslant \|a\|^n$$

$$\cdot \left(\frac{t^2 - r^2(M, M_0)}{4}\right)^{n-1} \frac{1}{[(n-1)!]^2 \, n} \quad (n=2, 3, \ldots). \tag{8}$$

Note now that the points of the region $\mathscr{D}(M_0, R)$ satisfy the inequality

$$\frac{t^2 - r^2(M, M_0)}{4} \leqslant R^2.$$

This leads to

$$\|v_n\| \leqslant \|a\| \, (\|a\| \; R^2)^{n-1} \frac{1}{[(n-1)!]^2 \cdot n}. \tag{9}$$

With inequality (9), it is trivial to estimate the norm of $v(M, M_0, t) - v_1(M, M_0, t)$. Indeed, setting $\lambda = 1$ in (3), we find

$$\|v - v_1\| \leqslant \sum_{n=2}^{\infty} \|v_n\| \leqslant \|a\|^2 \, R^2 \sum_{k=1}^{\infty} \frac{(\|a\| \; R^2)^{k-1}}{(k!)^2 \, (k+1)}. \tag{10}$$

From this estimate one can see that the norm of the difference of the exact solution of the problem and its linear approximation is small if the norm of $a(M)$ is small or the parameter R is small. Given an a priori estimate for $\|a\|$, which sometimes can be done on the basis of physical considerations, to each given positive number ε, it is easy to find R such that in the region $\mathscr{D}(M_0, R)$, $\|v - v_1\| \leqslant \varepsilon$. For fixed ε, the parameter R is larger the greater is the norm of the function $a(M)$ in C. Thus, if on some set belonging to $\mathscr{D}(M_0, R)$ we know the true solution $v(M, M_0, t)$, then we know also $v_1(M, M_0, t)$ to within ε.

From the foregoing, the following version of the inverse problem proves to be justified for application: Given the function $v_1(M, M_0, t)$ at the points of the plane $y=0$, find the function $a(M)$. We shall consider two particular cases of this problem.

Formulation 1. M_0 is a fixed point of the plane $y=0$ and the function

$$v_1(M, M_0, t)|_{y=0} = \varphi(x, t)$$

is known at all points belonging to the intersection of the four-dimensional region $\mathscr{D}(M_0, R)$ and the hyperplane $y=0$. Find $a(M)$ in the sphere $r(M, M_0) \leqslant R$.

Recalling that the formula

$$v_1(M, M_0, t) = \frac{1}{\pi [t^2 - r^2(M, M_0)]} \iint_{S(M, M_0, t)} r^2(P, M_0) \, a(P) \, d\omega \tag{11}$$

holds for $v_1(M, M_0, t)$, we arrive at the integral-geometric problem for the function $b(M)=r^2(M, M_0)\,a(M)$ considered in Sec. 1 of Chapter 1: Knowing the integrals over all ellipsoids of revolution lying in the sphere $r(M, M_0)\leqslant R$ with one focus fixed at M_0 and the other running over a point set of $y=0$, find the function $b(M)$. Since, obviously, $b(M)$ satisfies a Hölder condition in a neighbourhood of the point M_0, the conditions of the uniqueness theorem for this problem are satisfied, and we are justified in stating the following theorem.

Theorem 2.3. *The continuous function $a(M)$ in $r(M, M_0)\leqslant R$ is uniquely determined by assignment of the function $v_1(M, M_0, t)$ at the points of the hyperplane $y=0$ belonging to the region*

$$\mathscr{D}(M_0, R):\{r(M, M_0)\leqslant t\leqslant 2R-r(M, M_0)\}.$$

Consider now the following version of the inverse problem.

Formulation 2. The point M_0 runs over a point set of the plane $y=0$ lying inside a circle of radius ε with center at M_1 (ε is an arbitrarily small fixed positive number) and the function $v_1(M_0, M_0, t)$ is known for the times $0\leqslant t\leqslant T$; find the function $a(M)$.

Letting the point M approach M_0 in (11), we find

$$v_1(M_0, M_0, 2t)=\frac{1}{4\pi}\iint\limits_{S(M_0,t)} a(P)\,d\omega. \tag{12}$$

Here $S(M_0, t)$ is a sphere of radius t with center at M_0. Formulation 2 is equivalent to the integral geometric problem considered by R. Courant in [18]: Find a function from its spherical means when the centers of the spheres run over a point set of the hyperplane $y=0$. From the results for this problem follows the theorem.

Theorem 2.4. *The assignment of the function $v_1(M_0, M_0, t)$ in the region $\mathscr{D}(M_1, T)$: $\{r(M_0, M_1)\leqslant\varepsilon,\ M_0, M_1\in\{y=0\},\ 0\leqslant t\leqslant T\}$ uniquely determines function $a(M)$, belonging to the space C, in the sphere $r(M, M_1)\leqslant\frac{1}{2}T$.*

Corollary. *If the point M_0 varies over the entire hyperplane $y=0$, the function $a(M)$ is uniquely determined in the region $|y|\leqslant\frac{1}{2}T$.*

Formulation 2 can be successfully used when the differential equation has a more involved structure than (2.1). We can consider, for example, the following problem in x_1, x_2, x_3 space. The differential equation

$$u_{tt}=\sum_{i,j=1}^{3} a_{ij}u_{x_ix_j}+\sum_{i=1}^{3} b_iu_{x_i}+au+\delta(M-M_0, t) \tag{13}$$

is given, in which the coefficients a_{ij}, b_i are known and depend only on the variable $x_3 = y$. Find the function $a(M)$ if it has compact support in the region $\mathscr{D}\{0 \leqslant y \leqslant H\}$ given the solution $u(M_0, M_0, t)$ of the Cauchy problem for Eq. (13) with zero initial data and that M_0 varies over a point set of the plane $y=0$. In the linearized version this problem reduces to the integral-geometric problem where a linear combination of the integrals of the function $a(M)$ is known over the quasispheres $S(M_0, t)$ and the volumes bounding them [70]. In this case the quasispheres $S(M_0, t)$ and the arising weight functions are invariant to displacement along the plane $y=0$. The result will not be presented here as it would require a more detailed investigation of the integral-geometric problem than given in Sec. 7 of Chapter 1.

We note also that the methods used to consider the integral-geometric problems allow one to construct the function $a(M)$ in each case from the given data.

4. The Problem of Finding the Coefficients of the Lower Order Derivatives in a Second-Order Equation

I. Consider in x_1, x_2, x_3 space the Cauchy problem

$$\left.\begin{aligned} u_{tt} &= \mathrm{L}u + 4\pi\,\delta(x - x^0, t), \\ u|_{t<0} &\equiv 0, \end{aligned}\right\} \tag{1}$$

for the function $u(x, x^0, t)$, where $x^0 = (x_1^0, x_2^0, x_3^0)$ is a fixed point and parameter in the problem and L is the elliptic operator

$$\begin{aligned} \mathrm{L}u &\equiv \sum_{i,j=1}^{3} a_{ij}(x)\, u_{x_i x_j} + \sum_{i=1}^{3} b_i u_{x_i} + c(x)\, u, \\ &\sum_{i,j=1}^{3} a_{ij}(x)\, t_i t_j \geqslant \mu_0 \sum_{i=1}^{3} t_i^2, \ \mu_0 > 0 \end{aligned} \tag{2}$$

Let $\mathscr{D}$ be a region of x_1, x_2, x_3 space bounded by surface S (not necessarily closed). Consider the following inverse problem: The coefficients $a_{ij}(i, j = 1, 2, 3)$ of the operator L are known; the point x^0 runs over a point set $\mathfrak{M}$ of the surface S; the solution $u(x, x^0, t)$ of (1) is known for $x, x^0 \in \mathfrak{M}$ at times t lying in a neighbourhood of the value $t = \tau(x, x^0)$; find in $\mathscr{D}$ the coefficients of (1) for the lower order derivatives.

We show that the above problem can be reduced to an integral-geometric problem. Let the coefficients a_{ij} be such that the assumptions of Sec. 1 are fulfilled. Moreover, let $a_{ij} \in C^3$, $b_i \in C^2$ and $c \in C$. We shall use for problem (1) the information presented in Sec. 1 and we shall retain the previous notation. Using for the function $u(x, x^0, t)$ the representation (1.14) and the continuity of the function $v(x, x^0, t)$ we find

$$\sigma(x^0, x) = \lim_{t \to \tau(x, x^0) + 0} \int_0^t u(x, x^0, t)\, dt. \tag{3}$$

Hence, by the condition of the problem, $\sigma(x^0, x)$ can be viewed as a known function of the points x and x^0 of the set $\mathfrak{M}$. But then, if we use (1.12), we find that for any pair of points $x, x^0 \in \mathfrak{M}$ the following integrals are known:

$$\int_{\Gamma(x, x^0)} \left(\sum_{i=1}^{3} \tilde{b}_i(x) \frac{dx_i}{ds} \right) ds = f(x^0, x), \tag{4}$$

where

$$\tilde{b}_i(x) = \sum_{k=1}^{3} b_k(x)\, b_{ki}(x). \tag{5}$$

The function $f(x^0, x)$ can be computed in terms of the function $\sigma(x^0, x)$ and the coefficients a_{ij} $(i, j = 1, 2, 3)$. Since the geodesics $\Gamma(x, x^0)$ involve only the coefficients of the higher derivatives, they can be regarded as known. Thus we arrive at the problem of determining the functions $\tilde{b}_i(x)$ $(i = 1, 2, 3)$ in the region bounded by surface S from known integrals of a linear combination of these functions along the geodesics joining the points of the surface S. Of course, for the above problem to be meaningful, it is necessary that the geodesics $\Gamma(x, x^0)$ pass inside the region $\mathscr{D}$ and to behave regularly in a certain sense. This imposes additional restrictions on the coefficients a_{ij}.

Let us show that the problem of determining the coefficient $c(x)$ reduces also to a problem in integral geometry. Since we have obtained a problem for the coefficients b_i $(i = 1, 2, 3)$ in which the coefficient $c(x)$ is not involved, in deriving an equation for $c(x)$, we shall regard the coefficients $b_i(x)$ $(i = 1, 2, 3)$ known.

Consider (1.20). Letting $t \to \tau(x, x^0)$ and taking into account that the volume integral inside the quasi-ellipsoid vanishes, we find

$$v(x, x^0, \tau(x, x^0)) = \frac{1}{4\pi} \int_0^{2\pi} d\varphi \int_0^{\tau(x, x^0)} \Big\{ \mathrm{L}_\xi^* \sigma(\xi, x)\, \sigma(x^0, \xi) \left| \frac{\partial(\xi_1, \xi_2, \xi_3)}{\partial(t, \tau, \varphi)} \right|_{t = \tau(x, x^0)} \Big\} d\tau. \tag{6}$$

Note that in this formula $d\tau = ds$ is the element of arclength along geodesic $\Gamma(x, x^0)$ in the Riemannian metric (1.4). Since the coefficients a_{ij} and b_i are known, the function $\sigma(\xi, x)$ is also known and in (6) only the coefficient $c(x)$ in operator L* is unknown. Choosing explicitly the summand corresponding to the coefficient $c(x)$ in L* and introducing the notation

$$R(x, x^0, \xi) = \frac{1}{4\pi}\,\sigma(\xi, x)\,\sigma(x^0, \xi) \int_0^{2\pi} \left|\frac{\partial(\xi_1, \xi_2, \xi_3)}{\partial(t, \tau, \varphi)}\right|_{t=\tau(x, x^0)} d\varphi,$$

$$Q(x, x^0, \xi) = \frac{1}{4\pi}\,[(L_\xi^*\sigma(\xi, x) - c(\xi)\,\sigma(\xi, x))\,\sigma(x^0, \xi)] \tag{7}$$

$$\cdot \int_0^{2\pi} \left|\frac{\partial(\xi_1, \xi_2, \xi_3)}{\partial(t, \tau, \varphi)}\right|_{t=\tau(x, x^0)} d\varphi,$$

we obtain for the coefficient $c(x)$ the equation

$$\int_{\Gamma(x, x^0)} R(x, x^0, \xi)\,c(\xi)\,d\tau = v(x, x^0, \tau(x, x^0)) - \int_{\Gamma(x, x^0)} Q(x, x^0, \xi)\,d\tau. \tag{8}$$

Since the function $v(x, x^0, \tau(x, x^0))$ is known at the point set $\mathfrak{M}$ on the surface S, the expression on the right-hand side of formula (8) is a known function of pairs of points of $\mathfrak{M}$ and for the coefficient $c(x)$ we actually have an integral-geometric problem in the region $\mathscr{D}$. Note also that the Jacobian $\dfrac{\partial(\xi_1, \xi_2, \xi_3)}{\partial(t, \tau, \varphi)}$ in (7) can be expressed at the points of the geodesic $\Gamma(x, x^0)$ in terms of the curvatures of the normal cross-sections of the surfaces $\tau(\xi, x^0) = \tau$ and $\tau(\xi, x) = \tau(x, x^0) - \tau$. In the following, we shall denote these surfaces by τ_1 and τ_2 respectively. The unknown Jacobian could be calculated directly by introducing at a variable point ξ of the extremal $\Gamma(x, x^0)$ a cylindrical coordinate system ϱ, z, φ and choosing the tangent to the extremal to be the z axis ($\varrho = 0$). We prefer another approach which yields a formula analogous to (6) but one not containing the Jacobian. In this connection it will be noted that the integral in (6) is nothing more than the limit as $t \to \tau(x, x^0)$ of the first integral in (1.15). We transform this integral introducing the curvilinear coordinates τ, ϱ, φ at the point ξ in the following way. Consider the extremal $\Gamma(x, x^0)$ and a point on it lying at a distance τ (in the Riemannian metric (1.4)) from x^0. Draw through this point a plane orthogonal (in the Riemannian metric) to the tangent at the given point to the extremal $\Gamma(x, x^0)$ and introduce in this plane polar coordinates ϱ, φ with pole at the point of intersection of the extremal and the plane. At least for t close to $\tau(x, x^0)$, the coordinates τ, ϱ, φ uniquely determine the point ξ. Moreover,

$$dv_\xi = \frac{d\tau \cdot \varrho\, d\varrho\, d\varphi}{\dfrac{d\tau_1}{\partial v}},$$

where v is a unit external normal to the surface τ_1 computed at its point of intersection with the extremal. In the variables ϱ, τ, φ, the first integral in (1.15) can be written as

$$\frac{1}{4\pi}\iiint\limits_{\mathscr{D}(x,x^0,t)} \mathrm{L}_\xi^* \sigma(\xi, x)\cdot\sigma(x^0, \xi)\,\delta(t-\tau(\xi, x)-\tau(\xi, x^0))\,dv_\xi$$

$$= \frac{1}{4\pi}\int\limits_{\frac{1}{2}(t-\tau(x,x^0))}^{\frac{1}{2}(t+\tau(x,x^0))} d\tau \cdot \int\limits_0^{2\pi} d\varphi \int\limits_0^{\varrho^*(t,\tau,\varphi)} \mathrm{L}_\xi^*\sigma(\xi, x)\,\sigma(x^0, \xi)\frac{1}{\dfrac{\partial\tau_1}{\partial v}}$$

$$\cdot\,\delta(t-\tau(\xi, x)-\tau(\xi, x^0))\,\varrho\, d\varrho$$

$$= \frac{1}{4\pi}\int\limits_{\frac{1}{2}(t-\tau(x,x^0))}^{\frac{1}{2}(t+\tau(x,x^0))} d\tau \int\limits_0^{2\pi} d\varphi \Bigg\{\mathrm{L}_\xi^*\sigma(\xi, x)\cdot\sigma(x^0, \xi)$$

$$\times \frac{\varrho}{\dfrac{\partial\tau_1}{\partial v}\cdot\dfrac{\partial}{\partial\varrho}[\tau(\xi, x)+\tau(\xi, x^0)]}\Bigg\}_{\varrho=\varrho^*}.$$

In these formulas $\varrho^* = \varrho^*(t, \tau, \varphi)$ is the polar equation of the contour of the cross-section of the region $\mathscr{D}(x, x^0, t)$ in the above plane. Passing to the limit as $t \to \tau(x, x^0)$ in (1.15) and letting

$$k_1(\tau, \varphi) = \lim_{\varrho\to 0}\frac{1}{\varrho}\frac{\partial}{\partial\varrho}\tau(\xi, x^0),$$

$$k_2(\tau, \varphi) = \lim_{\varrho\to 0}\frac{1}{\varrho}\frac{\partial}{\partial\varrho}\tau(\xi, x),$$

we find

$$v(x, x^0, \tau(x, x^0)) = \frac{1}{4\pi}\int\limits_{\Gamma(x,x^0)} d\tau\,\frac{1}{\dfrac{\partial\tau_1}{\partial v}}\,\mathrm{L}_\xi^*\sigma(\xi, x)$$

$$\cdot\sigma(x^0,\xi)\int\limits_0^{2\pi}\frac{d\varphi}{k_1(\tau,\varphi)+k_2(\tau,\varphi)}.$$

It is obvious that the values $k_1(\tau,\varphi)$ and $k_2(\tau,\varphi)$ are the curvatures of the normal cross-sections of the surface τ_1 and τ_2, respectively. Comparing formula (6) with the one just obtained we find

$$R(x,x^0,\xi)=\frac{1}{4\pi}\frac{\sigma(x^0,\xi)\,\sigma(\xi,x)}{\dfrac{\partial\tau_1}{\partial v}}\int\limits_0^{2\pi}\frac{d\varphi}{k_1(\tau,\varphi)+k_2(\tau,\varphi)},\tag{7'}$$

$$Q(x,x^0,\xi)=\frac{1}{4\pi}\frac{[\mathrm{L}_\xi^*\sigma(\xi,x)-c(\xi)\,\sigma(\xi,x)]\,\sigma(x^0,\xi)}{\dfrac{\partial\tau_1}{\partial v}}$$

$$\int\limits_0^{2\pi}\frac{d\varphi}{k_1(\tau,\varphi)+k_2(\tau,\varphi)}.\tag{7''}$$

II. We now proceed to study (4) and (8). Since the general case of integral-geometric problems (for arbitrary curves) has not been treated so far we shall consider only the cases where the higher coefficients a_{ij} $(i,j=1,2,3)$ are such that the integral-geometric problems corresponding to (4) and (8) can be investigated by the method considered in Chapter 1. First we shall consider (4).

1) Let the coefficients a_{ij} depend on only one of the coordinates, for definiteness $x_3=y$, and let $\mathscr{D}$ be a region of x, y space, $x=(x_1,x_2)$, contained in $0\leqslant y\leqslant H$. Introduce for the coefficients a_{ij} the notation of Sec. 8 of Chapter 1 and assume that for any unit vector $\alpha=(\alpha_1,\alpha_2)$ the function $v_\alpha(y)$ has a positive derivative on the segment $[0,H]$. The physical meaning of this assumption, as will be clarified later in Sec. 6, is as follows: the propagation speed of disturbances increases monotonically with increasing y for any fixed direction $l=(\alpha_1,\alpha_2,0)$ of x, y space. A sufficient condition for the coefficients $a_{ij}(y)$ will be indicated there also by which to each pair of points in the region $\mathscr{D}$ there corresponds one geodesic joining them. This condition reduces to the fulfilment, on the segment $[0,H]$, of the inequality

$$\frac{d}{dy}\left[\frac{v_\alpha'(y)\sqrt{s(y)}}{\sqrt{v_\alpha^3(y)}}\right]\geqslant 0.\tag{9}$$

We shall regard inequality (9) to be satisfied by the coefficients a_{ij} as well because the integral equations of Sec. 1 are obtained on the assumption that there exists a one-to-one correspondence between pairs of points and geodesics connecting them.

It is currently convenient to take $\mathfrak{M}$ to be a point set of the plane $y=0$. We are led then to the following problem for the coefficients $\tilde{b}_i(x, y)$: Integrals (4) are known for the geodesics joining any pair of points $(x^0, 0)$ and $(x, 0)$ of the plane $y=0$; find $\tilde{b}_i(x, y)$ $(i=1, 2, 3)$. It will be shown now that the problem in this setting has no unique solution. First of all it will be noted that in this case the conditions of Theorem 1.7 a fortiori will not be fulfilled no matter what three-parameter families of curves $\Gamma(x, x^0)$ are taken. Indeed, at the vertex of any curve $\Gamma(x, x^0)$, the component dy/ds of the unit tangent vector $(dx_1/ds, dx_2/ds, dy/ds)$ (in the Riemannian metric (1.4)) vanishes and hence, the condition about the independence of the weight functions stated in Theorem 1.7 cannot be fulfilled. Of course, the non-uniqueness in the above problem is not a result yet of the non-fulfilment of the conditions of Theorem 1.7, because these conditions are only sufficient and not necessary for uniqueness. We now show that all the coefficients $\tilde{b}_i (i=1, 2, 3)$ cannot in fact be found from their integrals (4). For this purpose we transform formula (4) eliminating dy/ds from it. To this end, we observe that along the geodesic $\Gamma(x, x^0)$ joining the points $(x, 0)$ and $(x^0, 0)$, to each continuously differentiable function $P(x_1, x_2, y)$, the identity

$$P(x_1, x_2, 0)-P(x_1^0, x_2^0, 0)$$
$$=\int_{\Gamma(x, x^0)} \frac{dP}{ds}\, ds \equiv \int_{\Gamma(x, x^0)} \left(\frac{\partial P}{\partial x_1}\frac{dx_1}{ds}+\frac{\partial P}{\partial x_2}\frac{dx_2}{ds}+\frac{\partial P}{\partial y}\frac{dy}{ds}\right) ds$$

holds. Letting

$$P(x_1, x_2, y)=\int_0^y \tilde{b}_3(x_1, x_2, y)\, dy,$$

we find

$$\int_{\Gamma(x, x^0)} \tilde{b}_3 \frac{dy}{ds}\, ds$$
$$=-\int_{\Gamma(x, x^0)} \left\{\frac{dx_1}{ds}\frac{\partial}{\partial x_1}\int_0^y \tilde{b}_3(x_1, x_2, y)\, dy\right.$$

$$+\frac{dx_2}{ds}\frac{\partial}{\partial x_2}\int_0^y \tilde{b}_3(x_1, x_2, y)\,dy\Big\}\,ds.$$

By this last equality, (4) can be rewritten as

$$\int_{\Gamma(x,x^0)}\left[Q_1\frac{dx_1}{ds}+Q_2\frac{dx_2}{ds}\right]ds=f(x^0, x), \tag{10}$$

where

$$\begin{aligned} Q_1(x_1, x_2, y)&=\tilde{b}_1(x_1, x_2, y)-\frac{\partial}{\partial x_1}\int_0^y \tilde{b}_3(x_1, x_2, y)\,dy,\\ Q_2(x_1, x_2, y)&=\tilde{b}_2(x_1, x_2, y)-\frac{\partial}{\partial x_2}\int_0^y \tilde{b}_3(x_1, x_2, y)\,dy. \end{aligned} \tag{11}$$

From (10) it is clear that it is meaningful to talk only about determining the functions Q_1 and Q_2. It is not difficult to show under our hypothesis on the coefficients $a_{ij}(y)$ that the functions Q_1 and Q_2 are uniquely determined by the integrals (10) in the class of functions having compact support in $\mathscr{D}$. Indeed in Sec. 8 of Chapter 1 we established that when the function $v_\alpha'(y)$ is positive it is possible to choose from among the geodesics $\Gamma(x, x^0)$, three-parameter families of curves $\Gamma_\alpha(x^0, t)$ on which the integral-geometric problem has a unique solution. Let us take two such families corresponding to independent unit vectors α_1 and α_2. Let us show that the weight vectors $F_{\alpha_i}=(dx_1/ds, dx_2/ds)$ $(i=1, 2)$ constructed for the families $\Gamma_{\alpha_1}(x^0, t)$ and $\Gamma_{\alpha_2}(x^0, t)$ are independent in the sense of Theorem 1.7. From formula (8.12) of Chapter 1 it follows that the equality

$$F_\alpha(\eta)=\left.\frac{dx}{ds}\right|_{y=\eta}=zB^{-1}(\eta)=t\alpha B^{-1}(\eta);\qquad t=\frac{1}{\sqrt{v_\alpha(\eta)}},$$

is fulfilled at the vertex (ξ, η) of the geodesic $\Gamma_\alpha(x^0, t)$. Since the matrix $B^{-1}(\eta)$ is non-singular, to the linearly independent vectors α_1 and α_2 there correspond linearly independent vectors $\alpha_1 B^{-1}(\eta)$ and $\alpha_2 B^{-1}(\eta)$. Therefore $F_{\alpha_1}(\eta)$ and $F_{\alpha_2}(\eta)$ are linearly independent vectors. Thus in this case, referring to Theorem 1.7, one can assert the uniqueness of the functions Q_1 and Q_2 determined by integrals (10). From the above arguments one can also see that to construct the functions Q_1 and Q_2 one need not have all the information contained in (10). It suffices to dispose x and x^0 in the plane $y=0$ in such a way that only the integrals over the families of geodesics $\Gamma_{\alpha_1}(x^0, t)$ and $\Gamma_{\alpha_2}(x^0, t)$ are known. For this purpose it is sufficient to draw at each point $(x^0, 0)$ in the plane $y=0$ two "observation lines" corresponding to the values $z=\alpha_1 t$, $z=\alpha_2 t$, $v_\alpha^{-1/2}(H) \leqslant t \leqslant v_\alpha^{-1/2}(0)$, and to allow the point $(x, 0)$ to vary only over the set of "observation points".

Thus we have established the following theorem.

Theorem 2.5. *Let the coefficients a_{ij} $(i, j=1, 2, 3)$ of the operator* L *depend only on the variable $y=x_3$ and satisfy in the region $\mathscr{D}\{0 \leqslant y \leqslant H\}$ the conditions $v'_\alpha(y)>0$ and* (9) *for any vector $\alpha=(\alpha_1, \alpha_2)$, $|\alpha|=1$.*

If we know the solution of (1) *at the points of the plane $y=0$ in a neighbourhood of the arrival time of a disturbance from a point source to these points which also varies over a point set in the hyperplane $y=0$, then the functions $Q_1(x, y)$ and $Q_2(x, y)$ related to the coefficients of* (1) *by the expression* (11) *are uniquely determined in the class of functions having compact support in $\mathscr{D}$.*

We indicate some consequences of this result for the coefficients of (1). If the function $\tilde{b}_3(x, y)$ is known, then from formulas (11) one can find the functions $\tilde{b}_1(x, y)$ and $\tilde{b}_2(x, y)$ from which one can obtain by inverting formula (5) the functions $b_i(x, y)$ $(i=1, 2, 3)$. The functions $\tilde{b}_1(x, y)$ and $\tilde{b}_2(x, y)$ can also be found uniquely if $\tilde{b}_3$ is a function of y only. Then $\tilde{b}_i(x, y)=Q_i(x, y)$ $(i=1, 2)$. In this case the coefficients $b_i(x, y)$ $(i=1, 2, 3)$ are determined to within an arbitrary one-dimensional function $\tilde{b}_3(y)$. Expecially interesting here is the case where the operator L has no mixed derivatives: $a_{ij}\equiv 0$ for $i\neq j$. Then, even if we do not know the coefficient $b_3=b_3(y)$ we can nevertheless uniquely determine the coefficients $b_1(x, y)$, $b_2(x, y)$. But in general one can find in this setting only two relations (11) between the coefficients b_i $(i=1, 2, 3)$.

The above version of the inverse problem can be extended by considering a problem for the coefficients b_i $(i=1, 2, 3)$ in a region $\mathscr{D}_0 \subset \mathscr{D}$ bounded by plane $y=H$ and surface S defined by the explicit equation

$$y=\varphi(x_1, x_2),$$

where $\varphi(x_1, x_2)$ is a continuously differentiable function. It is natural to take $\mathfrak{M}$ to be a point set of the surface S. Let us show that this version easily reduces to the previous one.

Indeed, in solving (4) we can assume the functions $b_i(x, y)$ $(i=1, 2, 3)$ to vanish identically outside the region $\mathscr{D}_0$. Then, by continuing the curves $\Gamma(x, x^0)$ up to intersection with $y=0$ we arrive at the previous integral-geometric problem in the region $\mathscr{D}\{0 \leqslant y \leqslant H\}$. In this case the functions $Q_1(x, y)$ and $Q_2(x, y)$ have the form

$$Q_1(x_1, x_2, y)=\tilde{b}_1(x_1, x_2, y)-\frac{\partial}{\partial x_1}\int\limits_{\varphi(x_1, x_2)}^{y} \tilde{b}_3(x_1, x_2, y)\, dy,$$

$$Q_2(x_1, x_2, y)=\tilde{b}_2(x_1, x_2, y)-\frac{\partial}{\partial x_2}\int\limits_{\varphi(x_1, x_2)}^{y} \tilde{b}_3(x_1, x_2, y)\, dy.$$

If the function $\tilde{b}_3(x, y)$ is regarded as known in the region $\mathscr{D}$ the original formulation of the problem can be considered in any region $\mathscr{D}_0\subset\mathscr{D}$ bounded by a piecewise smooth surface S. The set $\mathfrak{M}$, naturally, can be assumed in this case coincident with a point set of the surface S. The problem of finding the coefficients $b_1(x, y)$, $b_2(x, y)$ in $\mathscr{D}_0$ reduces by an analogous procedure to the previous one to the integral-geometric problem considered in the strip $0\leqslant y\leqslant H$.

2) Now, let the coefficients a_{ij} $(i, j=1, 2, 3)$ of the operator L be dependent only on the distance to a fixed point. Let this point be the origin of x_1, x_2, x_3 coordinates. Consider the spherical coordinates r, θ, φ related to x_1, x_2, x_3 by

$$x_1=r\sin\theta\cos\varphi,\quad x_2=r\sin\theta\sin\varphi,\quad x_3=r\cos\theta. \tag{12}$$

The coefficients $a_{ij}\equiv a_{ij}(r)$ $(i, j=1, 2, 3)$. We consider the version of the inverse problem for the coefficients $b_i(x)$ $(i=1, 2, 3)$ in the layer $\mathscr{D}\{0<r_0\leqslant r\leqslant 1\}$ and we let $\mathfrak{M}$ be a point set of the unit sphere $r=1$. Denoting the spherical coordinates of the point $x\in\mathfrak{M}$ by θ, φ and those of the point $x_0\in\mathfrak{M}$ by θ_0, φ_0, we can write (4) as

$$\int\limits_{\Gamma(\theta, \varphi, \theta_0, \varphi_0)} \sum_{i=1}^{3} \tilde{b}_i(x)\, dx_i=f(\theta_0, \varphi_0, \theta, \varphi). \tag{13}$$

It is convenient to convert to spherical coordinates r, θ, φ in the integral. Doing this, we arrive at the formula

$$\int\limits_{\Gamma(\theta, \varphi, \theta_0, \varphi_0)} (\tilde{b}_r\, dr+\tilde{b}_\theta\, d\theta+\tilde{b}_\varphi\, d\varphi)=f(\theta_0, \varphi_0, \theta, \varphi), \tag{14}$$

in which $\tilde{b}_r$, $\tilde{b}_\theta$, $\tilde{b}_\varphi$ are components of the row matrix $\tilde{b}_q$, $q=(r, \theta, \varphi)$, given in terms of the row matrix $\tilde{b}=\|\tilde{b}_i\|$ $(i=1, 2, 3)$ by

$$\tilde{b}_q=\tilde{b}\cdot T_q. \tag{15}$$

Here T_q is the matrix whose rows are the partial derivatives of the variables $x_i (i=1, 2, 3)$ with respect to r, θ, φ, namely

$$T_q=\begin{pmatrix} \sin\theta\cos\varphi & r\cos\theta\cos\varphi & -r\sin\theta\sin\varphi \\ \sin\theta\sin\varphi & r\cos\theta\sin\varphi & r\sin\theta\cos\varphi \\ \cos\theta & -r\sin\theta & 0 \end{pmatrix}. \tag{16}$$

Note that in this case it is more natural to consider the operator L in spherical coordinates. In this case the first order portion of operator L reduces to

$$\sum_{i=1}^{3} b_i(x)\, u_{x_i} = b_r u_r + \frac{b_\theta}{r} u_\theta + \frac{b_\varphi}{r\sin\theta} u_\varphi, \tag{17}$$

where the components b_r, b_θ, b_φ of the row matrix b_q are computed in terms of the components of the row matrix $b=\|b_i\|$ $(i=1, 2, 3)$ by the formula analogous to (15),

$$b_q = b\ T_q.$$

From this we find

$$b = b_q T_q^{-1}, \tag{18}$$

where T_q^{-1} is the inverse of T_q. Recalling that by (5), $\tilde{b}=b\bar{B}$ where $\bar{B}$ is the inverse of $\bar{A}=\|a_{ij}\|$ $(i, j=1, 2, 3)$, we arrive at a formula relating the coefficients $\tilde{b}_r$, $\tilde{b}_\theta$, $\tilde{b}_\varphi$ to the coefficients b_r, b_θ, b_φ,

$$\tilde{b}_q = b_q (T_q^{-1} \bar{B}\ T_q). \tag{19}$$

We next transform formula (14) to a form analogous to (10). Introduce the functions

$$\left.\begin{aligned} Q_\theta(r, \theta, \varphi) &= \tilde{b}_\theta(r, \theta, \varphi) + \frac{\partial}{\partial\theta}\int_r^1 \tilde{b}_r(r, \theta, \varphi)\, dr, \\ Q_\varphi(r, \theta, \varphi) &= \tilde{b}_\varphi(r, \theta, \varphi) + \frac{\partial}{\partial\varphi}\int_r^1 \tilde{b}_r(r, \theta, \varphi)\, dr. \end{aligned}\right\} \tag{20}$$

After integration by parts formula (14) becomes

$$\int_{\Gamma(\theta, \varphi, \theta_0, \varphi_0)} Q_\theta\, d\theta + Q_\varphi\, d\varphi = f(\theta_0, \varphi_0, \theta, \varphi). \tag{21}$$

Therefore one can naturally hope to determine in $\mathscr{D}$ (if the geometry of the geodesics $\Gamma(\theta, \varphi, \theta_0, \varphi_0)$ is sufficiently good) the functions Q_θ and Q_φ. The integral-geometric problem arising for the functions Q_θ and Q_φ can be investigated by the techniques of Chapter 1. Note that the family of curves $\Gamma(\theta, \varphi, \theta_0, \varphi_0)$ in this case is invariant to rotation about the origin. However investigation of this problem for arbitrary $a_{ij}(r)$ would require an investigation analogous to that of Sec. 8 in Chapter 1: Under what conditions on the coefficients $a_{ij}(r)$ does the integral-geometric problem have a unique solution. Though this investigation does not involve any basic difficulties, we shall not present it here. We shall consider a particular case where the operator formed from the higher derivatives of the operator L is the Laplacian Δ with coefficient $n^{-2}(r)$. In this case the geodesics $\Gamma(\theta, \varphi, \theta_0, \varphi_0)$ are plane curves lying in the plane of the great circle passing through the points on the sphere $r=1$ having coordinates (θ, φ) and (θ_0, φ_0). The sufficient condition that one geodesic correspond to each pair of points in the region $\mathscr{D}$ has the form

$$(\ln n(r))'' \geqslant 0, \quad (0<r_0 \leqslant r \leqslant 1), \tag{22}$$

and the condition guaranteeing that through any point of the region $\mathscr{D}$ there passes a geodesic $\Gamma(\theta, \varphi, \theta_0, \varphi_0)$ with vertex at this point reduces to the fulfillment of the inequality

$$(rn(r))' > 0, \quad (0<r_0 \leqslant r \leqslant 1). \tag{23}$$

Conditions (22) and (23) are widely known in seismology (see, for example, [17], [23]) and therefore they will not be derived here.

Let us show that when conditions (22) and (23) are satisfied the functions Q_θ and Q_φ can be uniquely found from formula (21) in the region $\mathscr{D}\{0<r_0 \leqslant r \leqslant 1\}$. We make use of the results of Sec. 6 of Chapter 1. Consider the cross-sections of region $\mathscr{D}$ in meridianal planes. Then $d\varphi=0$ and for the function Q_θ in any such section we come to the problem considered in Chapter 1. Indeed, let the angle θ vary in such a cross-section from $-\pi$ to π and let $Q_\theta(r, -\theta, \varphi)=Q_\theta(r, \theta, \varphi+\pi)$. Denoting by (ϱ, α) the coordinates of the vertex of the curve $\Gamma(\theta, \varphi, \theta_0, \varphi_0)$, lying in the given cross-section, we obtain for the function $Q_\theta(r, \theta, \varphi)$ the equation

$$\sum_{j=1}^{2} \int_{\varrho}^{1} Q_\theta(r, \theta, \varphi) \frac{m(\varrho)}{\sqrt{m^2(r)-m^2(\varrho)}}\, dr = v(\varrho, \alpha, \varphi), \tag{24}$$

$$(0<r_0 \leqslant \varrho \leqslant 1, \quad -\pi \leqslant \alpha \leqslant \pi, \quad \varphi=\text{const}),$$

where

$$\theta_j = \alpha + (-1)^j \int_{\varrho}^{r} \frac{m(\varrho)\, dr}{\sqrt{m^2(r)-m^2(\varrho)}}, \quad m(r)=r\, n(r), \tag{25}$$

is the polar equation of the geodesic with the parameters ϱ, α.

Since $m'(r)>0(0<r_0\leqslant r\leqslant 1)$, equation (24) is of the same type as (6.13) of Chapter 1. Therefore in each cross-section of the region $\mathscr{D}$ in a meridianal plane one can uniquely determine a continuous function Q_θ. So, if we know the function $f(\theta_0, \varphi_0, \theta, \varphi)$ at points (θ, φ) and (θ_0, φ_0) on the circumferences passing through the pole of the system r, θ, φ, we can find the function Q_θ. We make use of the fact that we know the function $f(\theta_0, \varphi_0, \theta, \varphi)$ for any other such system of points (θ, φ), (θ_0, φ_0) related to the pole situated at an arbitrary point of the surface of the sphere $r=1$. For this purpose, consider, along with the x_1, x_2, x_3 coordinates, the system x_1', x_2', x_3' connected with the new point. For simplicity it will be assumed that the new pole is located at the point with spherical coordinates $(\theta_1, 0)$ (with respect to the system r, θ, φ). Let the x_1, x_2, x_3 coordinate system be obtained from x_1', x_2', x_3' by means of the orthogonal transformation with matrix

$$S=\begin{pmatrix} \cos\theta_1 & 0 & \sin\theta_1 \\ 0 & 1 & 0 \\ -\sin\theta_1 & 0 & \cos\theta_1 \end{pmatrix}. \tag{26}$$

We can associate with the x_1', x_2', x_3' system spherical coordinates r', θ', φ' $(r'=r)$ by formulas identical to (12). Introduce the notation $\tilde{b}_{q'}=\|\tilde{b}_{r'}, \tilde{b}_{\theta'}, \tilde{b}_{\varrho'}\|$. Then

$$\tilde{b}_{q'}=\tilde{b}\, S\, T_{q'}, \tag{27}$$

where $T_{q'}$ differs from matrix T_q in that the variables r, θ, φ are replaced by r', θ', φ'. Then, for the functions $Q_{\theta'}$ and $Q_{\varphi'}$ computed in terms of $\tilde{b}_{\theta'}, \tilde{b}_{\varphi'}$ by formulas identical to (20) we arrive at (21) from which, similarly to the foregoing, we can find $Q_{\theta'}$. Now it remains only to consider the question how $Q_{\theta'}$ is expressed in terms of Q_θ and Q_φ. To this end we make use of formula (27) for $\tilde{b}_{q'}$ and the coordinate transformation formulas from r', θ', φ' to r, θ, φ:

$$\cos\theta=\cos\theta_1\,\cos\theta'-\sin\theta_1\,\sin\theta'\,\cos\varphi',$$
$$\sin\theta e^{i\varphi}=\sin\theta_1\,\cos\theta'+\cos\theta_1\,\sin\theta'\,\cos\varphi'+i\,\sin\theta'\,\sin\varphi'.$$

Expressing the elements of matrix $T_{q'}$ in terms of the variables θ, φ we find

$$\begin{aligned}&\tilde{b}_{r'}=\tilde{b}_r,\\ &\sin\theta'\,\tilde{b}_{\theta'}=\tilde{b}_\theta(\sin\theta\,\cos\theta_1-\sin\theta_1\,\cos\theta\,\cos\varphi)+\tilde{b}_\varphi\,\frac{\sin\theta_1\,\sin\varphi}{\sin\theta}.\end{aligned} \tag{28}$$

On the other hand, using the first of these equalities, it is not difficult to verify that

$$\sin\theta'\,\frac{\partial\tilde{b}_{r'}}{\partial\theta'}=\frac{\partial\tilde{b}_r}{\partial\theta}\cdot(\sin\theta\;\cos\theta_1-\sin\theta_1\,\cos\theta\,\cos\varphi)+\frac{\partial\tilde{b}_r}{\partial\varphi}\,\frac{\sin\theta_1\,\sin\varphi}{\sin\theta}. \tag{29}$$

Thus, taking into account formulas (28) and (29), we find

$$\sin\theta' Q_{\theta'} = Q_\theta(\sin\theta\cos\theta_1 - \sin\theta_1\cos\theta\cos\varphi) + Q_\varphi\frac{\sin\theta_1\sin\varphi}{\sin\theta}. \quad (30)$$

Since the function Q_θ has already been found this implies that if $\sin\theta_1 \neq 0$, we can find Q_φ, as required. Thus, for determining the functions Q_θ, Q_φ, it is sufficient to have an observation system involving two different poles, i.e. two functions of the three variables.

In the case under consideration formula (19) has the form

$$\tilde{b}_q = n^2(r)\, b_q, \quad (19')$$

and hence, if the coefficient b_r is a known function in the region $\mathcal{D}$ or if it depends only on the variable r, one can find in our formulation the coefficients $b_\theta(r,\theta,\varphi)$, $b_\varphi(r,\theta,\varphi)$. This formulation of the problem certainly can also be extended to more complicated regions than $\mathcal{D}$.

We state the results obtained as a theorem.

Theorem 2.6. *Let the coefficients a_{ij} $(i,j=1,2,3)$ of operator* L *depend only on the distance from the origin of coordinates where $a_{ij}=(1/n^2(r))\,\delta_{ij}$ (δ_{ij} is the Kronecker delta) and let the function $n(r)$ satisfy conditions* (22) *and* (23). *Then if one knows the solution of* (1) *as a function of pairs of points x and x^0 on the sphere $r=1$ in a neighbourhood of the time $t=\tau(x,x^0)$, one can uniquely find in the region $\mathcal{D}\{0<r_0\leqslant r\leqslant 1\}$ the functions Q_θ and Q_φ, related by formulas* (17), (19′) *and* (20) *to the coefficients b_i $(i=1,2,3)$ of the operator* L.

The case where the coefficients of the higher order derivatives of the operator L are constant is within the scope of applicability of this theorem. Thus, by an appropriate orthogonal coordinate transformation and by scale changes along the x_1, x_2, x_3 axes, we can reduce the operator involving the second derivatives to the Laplacian with coefficient $n(r)\equiv 1$. Inequalities (22) and (23) are then fulfilled and we are under the conditions of applicability of the theorem proved above. It is necessary, of course, to take into account the fact that Q_θ and Q_φ will be changed into another form after the coordinate transformations.

III. We now proceed to consider Eq. (8) for $c(x)$. Here we have to first clarify the behaviour of the weight function $R(x,x^0,\xi)$. Generally this involves elaborate calculations. Therefore, we shall restrict ourselves to the inverse problem in which (1) has the form

$$n^2 u_{tt} = \Delta u + \sum_{i=1}^{3} b_i u_{x_i} + cu + 4\pi\delta(x-x^0,t). \quad (31)$$

Here Δ is the Laplace operator in the variables x_1, x_2, x_3. And in this latter case again we shall have to further restrict the problem by considering only coefficients $n(x)$ which lead to the integral-geometric problems of Chapter 1.

1) Let the coefficient n be a function only of the variable $x_3=y$. Consider the typical region $\mathscr{D}\{0\leqslant y\leqslant H\}$ (in this case) and the problem of determining the function $c(x)$ in $\mathscr{D}$ when the coefficients n and $b_i\,(i=1, 2, 3)$ are known. Again $\mathfrak{M}$ will be taken as a point set of the plane $y=0$. Since the function n defining the geometry of the geodesics depends only on the variable y, the family of geodesics $\Gamma(x, x^0)$ is clearly invariant to the displacement along the hyperplane $y=0$. It is important, however, to know also the behaviour of the weight function $R(x, x^0, \xi)$. Let us show that the function $n\cdot R(x, x^0, \xi)$ is independent of the variable ξ. The necessary calculations are as follows.

In the case of (31) the formula for the function $\sigma(\xi, x^0)$ has the form

$$\sigma^2(\xi, x^0)=\frac{n(x^0)\sin\theta_0\exp\displaystyle\int_{\Gamma(\xi,x^0)}\sum_{i=1}^{3} b_i(x)\,dx_i}{n(\xi)\left|\dfrac{\partial(\xi_1,\xi_2,\xi_3)}{\partial(s,\theta_0,\varphi_0)}\right|}, \tag{32}$$

where s is the usual (Euclidean) arclength along the extremal $\Gamma(\xi, x^0)$; θ_0 and φ_0 are the spherical angular coordinates defining the direction of the tangent to the extremal at the point x^0. Let us compute the Jacobian in the denominator of formula (32) for the case $n=n(y)$ under consideration. We make use of the existing axial symmetry of the quasispheres $\tau(\xi, x^0)=$ const. Consider the cylindrical coordinate system r, y, φ at the point x^0. The extremal $\Gamma(\xi, x^0)$ is a plane curve lying in the plane $\varphi=\varphi_0$. Its equation in cylindrical coordinates, according to formula (8.13) of Chapter 1, is

$$r=\int_0^y \frac{a\,|dy|}{\sqrt{n^2(y)-a^2}}. \tag{33}$$

Here a is constant along the ray $\Gamma(\xi, x^0)$, equal to $|z|$ in terms of Sec. 8, Chapter 1. The equation of the extremal $\Gamma(\xi, x^0)$ can also be written as

$$n(y)\sin\theta=a, \tag{34}$$

where θ is the angle between the y axis and the tangent to the extremal $\Gamma(\xi, x^0)$. The distance τ between the points ξ and x^0 in Riemannian metric

(1.4) is given by the formula

$$\tau = \int_0^y \frac{n^2(y)\,|dy|}{\sqrt{n^2(y)-a^2}}. \tag{35}$$

In this formula τ can be regarded as a composite function of r and y if we assume that $a=a(r, y)$ is found from formula (33) as a function of the variables r, y. Therefore

$$\begin{aligned}\frac{\partial(\xi_1, \xi_2, \xi_3)}{\partial(s, \theta_0, \varphi_0)} &= +\frac{\partial(\xi_1, \xi_2, \xi_3)}{\partial(r, y, \varphi_0)} \frac{\partial(r, y, \varphi_0)}{\partial(s, \theta_0, \varphi_0)}\\ &= -r\frac{\partial(r, y)}{\partial(\tau, \theta_0)}\frac{d\tau}{ds} = -r\; n(y) \frac{\partial(r, y)}{\partial(\tau, a)}\frac{da}{d\theta_0}\\ &= -rn(y)\, n(y^0) \cos\theta_0 \frac{1}{\dfrac{\partial(\tau, a)}{\partial(r, y)}}.\end{aligned}$$

Here we made use of the fact that along the ray $\Gamma(\xi, x^0)$, we have $d\tau/ds = n(y)$ and of formula (34) putting in it $y=y^0$ and $\theta=\theta_0$. To compute the Jacobian $\frac{\partial(\tau, a)}{\partial(r, y)}$ we use formulas (33) and (35). Differentiating formula (33) and assuming $y>0$ and $dy>0$ we find

$$\left.\begin{aligned}\frac{\partial a}{\partial r} &= \left[\int_0^y \frac{n^2(y)\,dy}{\sqrt{(n^2(y)-a^2)^3}}\right]^{-1},\\ \frac{\partial a}{\partial y} &= -\frac{a}{\sqrt{n^2(y)-a^2}}\left[\int_0^y \frac{n^2(y)\,dy}{\sqrt{(n^2(y)-a^2)^3}}\right]^{-1}.\end{aligned}\right\} \tag{36}$$

In a similar way, differentiating (35) and employing the formulas obtained for $\partial a/\partial r$ and $\partial a/\partial y$ we find

$$\left.\begin{aligned}\frac{\partial\tau}{\partial y} &= \frac{n^2(y)}{\sqrt{n^2(y)-a^2}} + \int_0^y \frac{n^2(y)\,dy}{\sqrt{(n^2(y)-a^2)^3}}\cdot a\frac{\partial a}{\partial y} = \sqrt{n^2(y)-a^2},\\ \frac{\partial\tau}{\partial r} &= \int_0^y \frac{n^2(y)\,dy}{\sqrt{(n^2(y)-a^2)^3}}\cdot a\frac{\partial a}{\partial r} = a.\end{aligned}\right\} \tag{37}$$

Now everything is ready for determining the desired Jacobian. Computing it, we obtain

$$\frac{\partial(\tau, a)}{\partial(r, y)} = -\frac{n^2(y)}{\sqrt{n^2(y)-a^2}}\left[\int_0^y \frac{n^2(y)\,dy}{\sqrt{(n^2(y)-a^2)^3}}\right]^{-1}.$$

Note that the integral in brackets is not always meaningful. We obtained it on the assumption that $dy > 0$ along the path of integration. In the general case, this integral should be replaced by an expression having meaning everywhere. To this end, we note that

$$\int_0^y \frac{n^2(y)\,dy}{\sqrt{[n^2(y)-a^2]^3}} = \frac{1}{a}\frac{\partial}{\partial a}\int_0^y \frac{n^2(y)\,dy}{\sqrt{n^2(y)-a^2}} = \frac{1}{a}\frac{\partial \tau}{\partial a}.$$

Using this and formula (34) we get finally

$$\begin{aligned}\left|\frac{\partial(\xi_1, \xi_2, \xi_3)}{\partial(s, \theta_0, \varphi_0)}\right| &= \frac{r\, n(y^0)\,|\cos\theta_0|}{n(y)\, a}\sqrt{n^2(y)-a^2}\,\frac{\partial \tau}{\partial a}\\ &= r\, n(y^0)\,|\cos\theta_0|\;|\cot\theta|\,\frac{\partial\tau}{\partial a}\end{aligned}$$

and formula (32) becomes

$$\sigma^2(\xi, x^0) = \frac{|\tan\theta_0 \tan\theta|}{r\dfrac{\partial\tau}{\partial a}}\exp\int_{\Gamma(\xi, x^0)}\sum_{i=1}^{3} b_i(x)\,dx_i. \tag{38}$$

To compute the function $R(x, x^0, \xi)$ we must know $\sigma(\xi, x)$ and $\sigma(x^0, \xi)$ at the points of the extremal $\Gamma(x, x^0)$. Denoting the angle between the y axis and the tangent to the extremal $\Gamma(x, x^0)$ at the point x by θ_1 and the cylindrical coordinate of point x by ϱ we find by formula (38) the following expression:

$$\left.\begin{aligned}\sigma^2(x^0, \xi) &= \frac{|\tan\theta_0 \tan\theta|}{r\dfrac{\partial\tau_1}{\partial a}}\exp\int_{\Gamma(\xi, x^0)}\sum_{i=1}^{3} b_i(x)\,dx_i,\\ \sigma^2(\xi, x) &= \frac{|\tan\theta_1 \tan\theta|}{(\varrho - r)\dfrac{\partial\tau_2}{\partial a}}\exp\int_{\Gamma(\xi, x)}\sum_{i=1}^{3} b_i(x)\,dx_i.\end{aligned}\right\} \tag{39}$$

Now we proceed to compute the curvature of the normal cross-sections of the surfaces τ_1 and τ_2. We shall make use of the familiar theorem of differential geometry which says that at any point of a surface there exist two perpendicular directions in the tangent plane to the surface in which the curvature of the normal cross-sections attains a maximum and minimum. In our case, since the surfaces τ_1 and τ_2 have axial symmetry, one of these directions lies in a plane passing through the y-axis and the other is orthogonal to it. By Euler's formula then,

$$k_i(\tau, \varphi) = k_i^1(\tau)\cos^2\varphi + k_i^2(\tau)\sin^2\varphi, \quad (i=1, 2), \tag{40}$$

where k_1^1 and k_1^2 are the principal curvatures of the normal cross-sections of the surface $\tau_i (i=1, 2)$. The angle φ is to be measured in the tangent plane from the direction which yields the curvature k_i^1. Let us first find the principal curvatures k_1^1 and k_2^1 corresponding to the surface τ_1. Since this surface has the y-axis as axis of symmetry, k_1^2 can be easily computed.

Consider the extremal $\Gamma(x, x^0)$. At its point of intersection with the surface τ_1 draw a tangent to it. This tangent is orthogonal to the surface τ_1 and crosses the y-axis. The curvature k_1^2 of the principal cross-section in the plane through the tangent and perpendicular to the plane containing the tangent and the y-axis is equal to the reciprocal of the length of the segment of the tangent contained between the y-axis and the surface τ_1. Taking into account the fact that the angle between the y-axis and the tangent to the extremal has been denoted by θ, we find

$$k_1^2(\tau) = \frac{\sin\theta}{r}. \tag{41}$$

We calculate now the curvature $k_1^1(\tau)$ of the cross-section of the surface τ_1 in the plane going through the y-axis or, in other words, the curvature of the curve whose equation is represented by formula (35) with $\tau = \text{const.}$ and a a function of the variables r, y implicitly determined by formula (33). Denoting the element of arclength of the curve (35) by dl, we find

$$k_1^1(\tau) = \frac{d\theta}{dl} = -\sin\theta\,\frac{d\theta}{dy}. \tag{42}$$

Differentiating (35) with respect to y keeping τ fixed and using formula (34), we obtain

$$0 = \frac{n^2(y)}{\pm\sqrt{n^2(y)-a^2}} + \frac{\partial\tau_1}{\partial a}\left[n'(y)\sin\theta + n(y)\cos\theta\,\frac{d\theta}{dy}\right]. \tag{43}$$

We have to take the $(+)$ sign in this if in moving along the extremal $dy>0$ and the $(-)$ sign if $dy<0$. Note that in either case

$$\pm\sqrt{n^2(y)-a^2} = n(y)\cos\theta.$$

Finding $d\theta/dy$ from formula (43) and substituting it into (42) we obtain

$$k_1^1(\tau)=\frac{n'(y)\sin^2\theta}{n(y)\cos\theta}+\frac{\sin\theta}{\cos^2\theta\,\dfrac{\partial\tau_1}{\partial a}}. \tag{44}$$

The formulas for the principal curvatures k_2^1 and k_2^2 of the normal cross-sections of the surface τ_2 are obtained from the respective formulas (41) and (44) by replacing θ by $\pi-\theta$ and r by $\varrho-r$. Thus

$$\left.\begin{aligned} k_2^2&=\frac{\sin\theta}{\varrho-r},\\ k_2^1&=-\frac{n'(y)\sin^2\theta}{n(y)\cos\theta}+\frac{\sin\theta}{\cos^2\theta\,\dfrac{\partial\tau_2}{\partial a}}.\end{aligned}\right\} \tag{45}$$

We calculate now $R(x, x^0, \xi)$ from formula (7′). First of all, using (40), we find

$$\begin{aligned}\int_0^{2\pi}\frac{d\varphi}{k_1(\tau,\varphi)+k_2(\tau,\varphi)}&=\int_0^{2\pi}\frac{d\varphi}{[k_1^1(\tau)+k_2^1(\tau)]\cos^2\varphi+[k_1^2(\tau)+k_2^2(\tau)]\sin^2\varphi}\\ &=\frac{2\pi}{\sqrt{[k_1^1(\tau)+k_2^1(\tau)]\,[k_1^2(\tau)+k_2^2(\tau)]}}\\ &=\frac{2\pi|\cot\theta|\sqrt{r(\varrho-r)\dfrac{\partial\tau_1}{\partial a}\dfrac{\partial\tau_2}{\partial a}}}{\sqrt{\varrho\left(\dfrac{\partial\tau_1}{\partial a}+\dfrac{\partial\tau_2}{\partial a}\right)}}.\end{aligned} \tag{46}$$

Note that by definition of the surfaces τ_1 and τ_2, $\tau_1+\tau_2=\tau(x, x^0)$ at points of the extremal $\Gamma(x, x^0)$ which implies

$$\frac{\partial\tau_1}{\partial a}+\frac{\partial\tau_2}{\partial a}=\frac{\partial}{\partial a}\tau(x, x^0).$$

We substitute the expressions for $\sigma(\xi, x)$ and $\sigma(x^0, \xi)$ from (39) and the expression found for the integral (46) into formula (7′). Then, using the fact that for (31) the direction of the normal ν to the surface τ_1 coincides with that of the tangent to the extremal and, hence, that

$$\frac{d\tau_1}{d\nu}=n(y),$$

we find

$$R(x, x^0, \xi) = \frac{1}{2n(y)} \sqrt{\frac{|\tan\theta_0 \tan\theta_1|}{\varrho \frac{\partial}{\partial a}\tau(x, x^0)}} \exp \tfrac{1}{2} \int\limits_{\Gamma(x, x^0)} \sum_{i=1}^{3} b_i(x)\, dx_i \equiv \frac{R^0(x, x^0)}{n(y)}. \tag{47}$$

Thus, we have found that the function $R(x, x^0, \xi)$ depends on the variable ξ just through the function $n(y)$. We divide both sides of (8) by $R^0(x, x^0)$ and note that $d\tau/n(y) = ds$ along the extremal, where ds is the element of Euclidean arclength along the extremal. Then for the function $c(x)$ in the region $\mathscr{D}\{0 \leqslant y \leqslant H\}$ we obtain a typical integral-geometric problem: The integrals

$$\int\limits_{\Gamma(x, x^0)} c(\xi)\, ds = f(x, x^0)$$

are known for points x and x^0 belonging to the plane $y=0$; find $c(x)$. For uniqueness in this problem, as follows from the results of Sec. 8, it suffices that $n'(y) < 0$ for $0 \leqslant y \leqslant H$. Recalling further that for the uniqueness of the correspondence of an extremal to a pair of points x, x^0 joining these points it suffices that $(\ln n(y))'' \geqslant 0$, we come to the following theorem.

Theorem 2.7. *Let the coefficients n and b_i $(i=1, 2, 3)$ of (31) be given with n dependent only on the variable $x_3 = y$ and let $n'(y) < 0$ and $(\ln n(y))'' \geqslant 0$, $0 \leqslant y \leqslant H$. Then the coefficient $c(x)$ is uniquely determined in the class of functions having compact support in $\mathscr{D}\{0 \leqslant y \leqslant H\}$ if we know the solution $u(x, x^0, t)$ of (31) at $x, x^0 \in \mathfrak{M}\{y=0\}$ at times t lying in a neighbourhood of the value $t = \tau(x, x^0)$.*

2) Consider now another case of (31) where n is a function only of the distance r to a fixed point which will be regarded as the origin of the coordinate system. In this case, similarly to the foregoing, the evaluation of $R(x, x^0, \xi)$ is easy to carry out. We shall not present the details but shall indicate only the basic formulas. The extremal $\Gamma(x, x^0)$ is now a plane curve. Draw a plane through the origin containing the curve $\Gamma(x, x^0)$ and introduce polar coordinates r, ψ in this plane with the pole of the system the origin and the polar axis the straight line through the origin and the point x^0. Denote the polar coordinates of the points x and x^0 by $(r_0, 0)$ and (r_1, ψ_1) respectively. Then the formulas

$$\psi = \int\limits_{r_0}^{r} \frac{a\,|dr|}{r\sqrt{m^2(r) - a^2}},$$

$$\tau = \int_{r_0}^{r} \frac{m^2(y)\,|dr|}{r\sqrt{m^2(r)-a^2}} \tag{48}$$

are the analogues of (33) and (35) yielding the equation of the extremal $\Gamma(x, x^0)$ and its arclength in the Riemannian metric. In the above formulas $m(r)$ is given in terms of $n(r)$ by the formula

$$m(r) = r\; n(r),$$

and a is constant along the extremal $\Gamma(x, x^0)$. Denote by θ the angle between the tangent to the extremal $\Gamma(x, x^0)$ and the radius vector drawn from the origin to the point of tangency. Then the equation of the extremal can also be written in the form

$$m(r)\sin\theta = a. \tag{49}$$

The formulas for $\sigma(x^0, \xi)$ and $\sigma(\xi, x)$ along the extremal are analogous to (39) and have the form

$$\left.\begin{aligned} \sigma^2(x^0, \xi) &= \frac{|\tan\theta_0 \tan\theta|}{r_0 r \sin\psi \dfrac{\partial\tau_1}{\partial a}} \exp \int_{\Gamma(x_0, \xi)} \sum_{i=1}^{3} b_i(x)\,dx_i, \\ \sigma^2(\xi, x) &= \frac{|\tan\theta_1 \tan\theta|}{r_1 r |\sin(\psi_1 - \psi)| \dfrac{\partial\tau_2}{\partial a}} \exp \int_{\Gamma(\xi, x)} \sum_{i=1}^{3} b_i(x)\,dx_i, \end{aligned}\right\} \tag{50}$$

where θ_0 and θ_1 are the values of θ at points x and x_0 of $\Gamma(x, x^0)$, respectively.

The formulas for the principal curvatures of the surfaces τ_1 and τ_2 have the form

$$\left.\begin{aligned} k^2 &= \frac{\sin(\theta-\psi)}{r\sin\psi}, \qquad k_2^2 = \frac{\sin(\theta+\psi_1-\psi)}{r\sin(\psi_1-\psi)}, \\ k_1^1 &= \frac{m'(r)\sin^2\theta}{m(r)\cos\theta} + \frac{\sin\theta}{r\cos^2\theta \dfrac{\partial\tau_1}{\partial a}}, \\ k_2^1 &= -\frac{m'(r)\sin^2\theta}{m(r)\cos\theta} + \frac{\sin\theta}{r\cos^2\theta \dfrac{\partial\tau_2}{\partial a}}. \end{aligned}\right\} \tag{51}$$

Substituting the expressions for $\sigma(x^0, \xi)$ and $\sigma(\xi, x)$ and the principal curvatures from (50) and (51) into formula (7′), we find

$$R(x, x^0, \xi) = \frac{1}{2rn(r)} \sqrt{\frac{|\tan\theta_0 \tan\theta_1|}{r_0 r_1 \sin\psi_1 \frac{\partial}{\partial a}\tau(x, x^0)}} \cdot \exp\frac{1}{2} \int\limits_{\Gamma(x, x^0)} \sum_{i=1}^{3} b_i(x)\, dx_i \equiv \frac{R^0(x, x^0)}{rn(r)}. \tag{52}$$

Dividing, as before, both sides of (8) by $R^0(x, x^0)$ and taking into account that $r = |\xi|$ and $d\tau = n(r)\, ds$, we find for $c(x)$ the equation

$$\int\limits_{\Gamma(x, x^0)} \frac{c(\xi)}{|\xi|}\, ds = f(x, x^0). \tag{53}$$

Consider now the region $\mathscr{D}\{0 < r_0 \leqslant r \leqslant 1\}$ and let $\mathfrak{M}$ be a point set on the sphere $r = 1$. Then, if $m'(r) > 0$, $(\ln m(r))'' \geqslant 0$ for $r_0 \leqslant r \leqslant 1$, the integral-geometric problem (53) has a unique solution in the class of continuous functions and we obtain a theorem analogous to Theorem 2.7.

Theorem 2.8. *Let the coefficients n and b_i $(i = 1, 2, 3)$ be given with n dependent only on the distance r to the origin of coordinates and $(r\, n(r))' > 0$, $(\ln r\, n(r))'' \geqslant 0$, for $0 < r_0 \leqslant r \leqslant 1$. Then the coefficient $c(x)$ is uniquely determined in the region $\mathscr{D}\{0 < r_0 \leqslant r \leqslant 1\}$ in the class of continuous functions by the solution $u(x, x^0, t)$ of (31) known for pairs of points x, x^0 belonging to the surface of the sphere $r = 1$ at time t close to $\tau(x, x^0)$.*

We note further that in case $n \equiv 1$, $b_i \equiv 0$ $(i = 1, 2, 3)$, equation (8) has an especially simple structure.

In this case

$$\sigma(\xi, x) = \frac{1}{|\xi - x|}, \qquad \sigma(x^0, \xi) = \frac{1}{|\xi - x^0|},$$

$$k_1^1 = k_1^2 = \frac{1}{|\xi - x^0|}, \qquad k_2^1 = k_2^2 = \frac{1}{|\xi - x|}.$$

Taking into account that $L^* = \Delta$ and $\Delta(1/|\xi - x|) = 0$, we find

$$\frac{1}{2|x - x^0|} \int\limits_{\Gamma(x, x^0)} c(\xi)\, ds = v(x, x^0, \tau(x, x^0)). \tag{54}$$

Here $\Gamma(x, x^0)$ is the straight line joining the points x and x^0.

IV. To conclude this section, we shall briefly consider one more question involving the determining of coefficients of the derivatives. Namely, instead of (1) we shall consider a more general equation

$$u_{tt} = \mathrm{L}u + hu_t + 4\pi\,\delta(x - x^0, t), \tag{55}$$

where L is the operator given by formula (2). The Cauchy problem for this equation with zero initial data at $t=0$ reduces to solving (13) with $\sigma(\xi, x)$ of a slightly different form than (12). The difference in $\sigma(\xi, x)$ is that for (55) the function h must be added to the integrand in formula (12). Therefore referring to the version of the inverse problem for (55) stated at the beginning of the section, we have instead of (4) the equation

$$\int\limits_{\Gamma(x,\, x^0)} \left[\sum_{i=1}^{3} \tilde{b}_i(x) \frac{dx_i}{ds} + h\right] ds = f(x^0, x). \tag{56}$$

The integral-geometric problem arising for the functions $\tilde{b}_i$ and h is easily investigated in the same cases as problem (4). We shall do this, for example, for the case where the coefficients a_{ij} depend only on the variable $x_3 = y$. Then (56) reduces to

$$\int\limits_{\Gamma(x,\, x^0)} \left[Q_1 \frac{dx_1}{ds} + Q_2 \frac{dx_2}{ds} + h\right] ds = f(x^0, x), \tag{57}$$

where Q_1 and Q_2 are given by formulas (11). It is easy to show that the problem of determining Q_1, Q_2 and h from (57) has a unique solution in the class of functions with compact support in the region $\mathscr{D}\{0 \leqslant y \leqslant H\}$ if $v'_\alpha(y) > 0$ for $y \in [0, H]$, and the points x, x^0 run over the plane $y=0$. Indeed, reasoning as before, we choose three families of curves $\Gamma_\alpha(x^0, t)$ corresponding to three pairwise independent unit vectors α_i $(i=1, 2, 3)$. Then the corresponding vectors

$$F_{\alpha_i} = \left(\frac{dx_1}{ds}, \frac{dx_2}{ds}\right)$$

computed at the vertices of the curves $\Gamma_{\alpha_i}(x^0, t)$ will also be pairwise independent unit vectors (in the Riemannian metric, a fact already shown). Then the vectors

$$\Phi_{\alpha_i} = \left(\frac{dx_1}{ds}, \frac{dx_2}{ds}, 1\right)$$

computed at the vertices of the curves $\Gamma_{\alpha_i}(x^0, t)$ will have the following structure: their projections on the y-axis are constant and the projections on the plane $y=0$ are piecewise independent vectors of "unit" length. From this it is clear that the vectors Φ_{α_i} $(i=1, 2, 3)$ cannot lie in the same plane; hence, they are linearly independent. But then we have the conditions of applicability of Theorem 1.7 and we can assert the uniqueness of the solution of (57) in the above mentioned class.

Thus we have established the following theorem.

Theorem 2.9. *Let the coefficients* a_{ij} *of operator* L *be dependent only on the variable* $x_3 = y$ *and in the region* $\mathscr{D}\{0 \leqslant y \leqslant H\}$ *let* $v'_\alpha(y) > 0$ *and let condition* (9) *hold. Then the functions* Q_1, Q_2, *related to the coefficients of* L *by relations* (11), *and the coefficient h are uniquely determined in the class of functions having compact support in* $\mathscr{D}$ *by the solution of the Cauchy problem for* (55) *with zero initial data at the points* x, x^0 *of the plane* $y = 0$ *for times t lying in a neighbourhood of the value* $t = \tau(x, x^0)$.

5. Linearized Inverse Kinematic Problem for the Wave Equation in Variable Isotropic Media

I. Consider in $x_1, \ldots, x_m$ space ($m \geqslant 2$) the wave process described by the equation

$$n^2 u_{tt} = \Delta u + \delta(x - x^0, t) \tag{1}$$

under the initial conditions

$$u\,|_{t<0} \equiv 0. \tag{2}$$

Here Δ is the Laplacian operator in $x_1, \ldots, x_m$; n is a twice continuously differentiable function of $x = (x_1, \ldots, x_m)$ subject to the condition $0 < n_0 \leqslant n(x) \leqslant N_0$. Physically, $n(x)$ is the refractive index of the medium comprising $x_1, \ldots, x_m$ space while the function $1/n(x)$ is the disturbance propagation speed in the medium.

In many applications, in particular in geophysics, a problem is to determine the function $1/n(x)$ inside a given region $\mathscr{D}$ if some functionals of the solution of problem (1), (2) are known at the points of the surface S bounding this region. In [1], [3], [4], [7], [28], [29], [34], [47] and [84] various aspects of the problem have been studied. Most attention in the above works has been paid to the so-called inverse kinematic problem for (1) the essense of which is as follows. It is known that disturbances caused by a point source at $x^0 = (x_1^0, \ldots, x_m^0)$ propagate from this point along the geodesics $\Gamma(x, x^0)$ in the metric

$$ds = n(x)\,|dx|, \qquad |dx| = \sqrt{\sum_{i=1}^{m} dx_i^2}. \tag{3}$$

Moreover, the travel time $\tau(x, x^0)$ of a disturbance from the point x^0 to x

is equal to the "length" of the geodesic $\Gamma(x, x^0)$, namely,

$$\tau(x, x^0) = \int_{\Gamma(x, x^0)} n(x)\, |dx| . \tag{4}$$

Suppose the function $n(x)$ is such that the geodesics $\Gamma(x, x^0)$ proceed inside the region $\mathscr{D}$ when $x, x^0 \in S$.Then the following formulation is natural: $\tau(x, x^0)$ is known as a function of x andx_0 on the surface S; find the function $n(x)$ in $\mathscr{D}$. This formulation is equivalent to the following geometric problem: The distance between points of the surface S is known in the Riemannian metric (3); find this metric.

Note also that for fixed x^0 the function $\tau(x, x^0)$ satisfies the eiconal equation

$$|\operatorname{grad}_x \tau(x, x^0)| = n(x). \tag{5}$$

The inverse kinematic problem was first posed and solved in [34] and [84] and became the theoretical basis for the first attempts by seismology to study the deep structure of the Earth. The initial data for this was the seismological data observed in earthquakes, or to be more exact, the first arrival times of seismic waves registered at the seismological stations. The cited works deal with the solution of the problem of determining a function n depending only on one coordinate, e.g. the coordinate x_m, in half-space $x_m \geqslant 0$ from the travel time of disturbances from a fixed point of the hyperplane $x_m = 0$ to an arbitrary point of the same plane. Simultaneously it was noted that the solution is adequate only in the class of $n(x_m)$ that are monotonic functions of x_m. When the function $n(x_m)$ is not monotonic, the solution of the problem in this setting is not unique. Later the nature of nonuniqueness of the problem was studied in [28]. Some results for the non-one-dimensional problem are contained in [6] and [7].

II. Let us consider linearization of the inverse kinematic problem. For this purpose it is convenient to use (5). Suppose that the function $n(x)$ can be represented in the form

$$n(x) = n_0(x) + n_1(x), \tag{6}$$

where the function $n_0(x)$ is known and $n_1(x)$ is small in the C^2 norm as compared with the norm of $n_0(x)$. Introduce the perturbation parameter λ by

$$n(x) = n_0(x) + \lambda n_1(x) \tag{7}$$

and represent the function $\tau(x, x^0)$ as a series with respect to the perturbation parameter

$$\tau(x, x^0) = \sum_{n=0}^{\infty} \lambda^n \tau_n(x, x^0). \tag{8}$$

We then substitute the expressions for $n(x)$ and $\tau(x, x^0)$ into formula (5) and we equate like powers of λ. This yields for the functions $\tau_0(x, x^0)$ and $\tau_1(x,x^0)$ the equations

$$|\mathrm{grad}_x \tau_0(x, x^0)| = n_0(x), \tag{9}$$

$$(\mathrm{grad}_x \tau_0(x, x^0), \mathrm{grad}_x \tau_1(x, x^0)) = n_0(x)\, n_1(x). \tag{10}$$

The equations for any $\tau_n(x, x^0)$ are also easily written out but we shall not do this. We restrict ourselves to the linear approximation for $\tau(x, x^0)$ in λ only and, hence, in $n_1(x)$ also. Formula (8) can then be written as

$$\tau(x, x^0) = \tau_0(x, x^0) + \tau_1(x, x^0) + \varepsilon(x,x^0), \tag{11}$$

where $\varepsilon(x, x^0)$ tends to zero as $\|n_1\|_{C^2} \to 0$ and is of second order in comparison with $\tau_1(x, x^0)$.

One can see from formula (9) that $\tau_0(x, x^0)$ is the travel time of the signal in the metric associated with $n_0(x)$. As the function $n_0(x)$ is known, the function $\tau_0(x, x^0)$ can also be regarded as known. Formula (10) can be reduced to a more suitable form. For this it will be recalled that $\mathrm{grad}_x \tau_0(x, x^0)$ is directed along the tangent to the curve $\Gamma_0(x, x^0)$, the geodesic in metric (3) (for $n = n_0(x)$) joining the points x and x^0. Dividing by $n_0(x)$, we write formula (10) in the form

$$\sum_{i=1}^{m} \frac{\partial \tau_1(x, x^0)}{\partial x_i} \cos(l, x_i) = n_1(x), \tag{12}$$

where l is the direction of the tangent to the geodesic $\Gamma_0(x, x^0)$. In the last formula the expression on the left hand side is the derivative in the direction l. Integrating (12) along the geodesic $\Gamma_0(x, x^0)$ and taking into account the above, we arrive at the formula

$$\tau_1(x, x^0) = \int_{\Gamma_0(x, x^0)} n_1(x)\, |dx|. \tag{13}$$

Formula (13) is basic to the subsequent discussion. If the function $\tau(x, x^0)$ is given on some set, then, using formula (11), we can regard the function $\tau_1(x, x^0)$ to be known on the same set with error $\varepsilon(x, x^0)$. Therefore the inverse kinematic problem can be stated as follows: Find function $n_1(x)$ inside a region $\mathscr{D}$ if $\tau_1(x, x^0)$ is a known function of x and x^0 on the surface S bounding $\mathscr{D}$. Since the curves $\Gamma_0(x, x^0)$ are known in this case, this is an integral-geometric problem. Whenever the function $n_0(x)$ is such that the corresponding problem can be investigated by the methods of Chapter 1, it is possible to formulate certain conclusions concerning the linearized inverse kinematic problem.

III. Let the function $n_0(x)$ be dependent only on one variable, say x_m, which for convenience will be denoted by y, all the other variables $x_1, \ldots, x_{m-1}$ remain as before if we set $m-1=n$. The coordinates of a variable point $x=(x_1, \ldots, x_m)$ will be denoted below by (x, y), where $x=(x_1, \ldots, x_n)$. Consider the region $\mathscr{D}\{0 \leqslant y \leqslant H\}$ of x, y space. We pose the problem (typical of Chapter 1): The function $\tau_1(x, y, x^0, y^0)$ is a known function of pairs of points in the hyperplane $y=0$; find the function $n_1(x, y)$ in the region $\mathscr{D}$. In Sec. 8 of Chapter 1 we found that for the solution of this problem to be unique it is sufficient that the metric coefficients generating the curves $\Gamma_0(x, y, x^0, y^0)$ satisfy monotonicity conditions. We can easily show by direct verification that in this case such conditions reduce to the requirement that the derivative of $n_0(y)$ be negative in the interval $[0, H]$,

$$n_0'(y) < 0, \quad (0 \leqslant y \leqslant H). \tag{14}$$

Note that here the $\Gamma_0(x, y, x^0, y^0)$ are plane curves; therefore one can solve the corresponding integral-geometric problem independently in separate two-dimensional cross-sections of the region $\mathscr{D}$ in planes going through the y-axis and an arbitrary straight line of the hyperplane $y=0$ through the origin. In this connection to construct the function $n_1(x, y)$ in such cross section it is sufficient to know $\tau_1(x, y, x^0, y^0)$ as a function of pairs of points of the straight line lying in the hyperplane $y=0$. Thus to find the function $n_1(x, y)$ in the whole region $\mathscr{D}$ it is sufficient to know the function $\tau_1(x, y, x^0, y^0)$ as the points (x, y) and (x^0, y^0) run over the totality of straight lines of the hyperplane $y=0$ emanating from one point. This information, obviously, represents a function of the $(n+1)$ st variable, i.e. its dimension coincides with that of the function $n_1(x, y)$ to be determined.

Summarizing the above, we obtain the following theorem.

Theorem 2.10. *Let the function $n(x, y) \in C^2$ and be representable in $\mathscr{D}\{0 \leqslant y \leqslant H\}$ by*

$$n(x, y) = n_0(y) + n_1(x, y),$$

where $n_0(y) \in C^2$ is a known function with negative values on the segment $[0, H]$ and $n_1(x, y)$ is a small function in the norm of C^2. Then the function $n_1(x, y)$ is uniquely determined in the class of functions having compact support in $\mathscr{D}$, in terms of the function $\tau_1(x, y, x^0, y^0)$, where (x, y), $(x^0, y^0) \in \mathfrak{M}\{y=0\}$.

Remark. The above theorem implies a similar uniqueness theorem for more general regions than $\mathscr{D}\{0 \leqslant y \leqslant H\}$ considered by us. Indeed, let $\mathscr{D}_0$ be a finite region bounded by a piecewise smooth closed surface S, and let the function $\tau_1(x, y, x^0, y^0)$ be specified on a point set of S. Then placing $\mathscr{D}_0$ in the strip $0 \leqslant y \leqslant H$ and extending the definition of the function $n_1(x, y)$ to be zero outside $\mathscr{D}_0$, we come to the problem already studied. It is true, $n_1(x, y)$ will now be a piecewise continuous function; however, it is clear that the method employed by us in Chapter 1 is extendable to this class of functions. From the above analysis it follows also that it is sufficient to specify the function $\tau_1(x, y, x^0, y^0)$ on the point set belonging to the cross-sections of the surface S by the $(n-1)$-parametric family of two-dimensional planes going through the y-axis.

Practically speaking, one can find the function $n_1(x, y)$ using the algorithm of Chapter 1. Moreover, it is advantageous to employ it directly to each plane cross-section. In this case one can regard $n_1(x, y)$ simply as a function of the two variables x and y. The equation of the Fourier transform $n_{1\lambda}(y)$ of the function $n_1(x, y)$ corresponding to (8.28) of Chapter 1 has the form

$$\int_0^{\eta} n_{1\lambda}(y)\, K_\lambda(y, \eta)\, dy = \tau_{1\lambda}(\eta), \quad (0 \leqslant t \leqslant H) \tag{15}$$

where

$$K_\lambda(y, \eta) = \frac{2n_0(y)}{\sqrt{n_0^2(y) - n_0^2(\eta)}} \cos \lambda \int_y^{\eta} \frac{n_0(\eta)}{\sqrt{n_0^2(y) - n_0^2(\eta)}}\, dy. \tag{16}$$

In formula (15), $\tau_{1\lambda}(\eta)$ is the Fourier transform of $\tau_1(\xi, \eta) = \tau_1(x, 0, x^0, 0)$ and ξ and η are related to x and x^0 by

$$\begin{aligned} \frac{x + x^0}{2} &= \xi, \\ \frac{|x - x^0|}{2} &= \int_0^{\eta} \frac{n_0(\eta)\, dy}{\sqrt{n_0^2(y) - n_0^2(\eta)}}. \end{aligned} \tag{17}$$

It is not difficult to see that ξ and η are the vertex coordinates of the plane curve $\Gamma_0(x, 0, x^0, 0)$. The second relation in (17) establishes that to each value of n there corresponds one value of $\frac{1}{2}|x - x^0|$. It is not difficult to show that when $(\ln n_0(y))'' \geqslant 0$ the inverse correspondence will also be single-valued. If the inequality is not satisfied, there may be more than one curve joining a pair of points $(x, 0)$, $(x^0, 0)$. Then, according to (15) we must know the function $\tau_1(x, 0, x^0, 0)$ for each of these curves because different values of the parameter η correspond to them. It is advantageous then to consider only functions $n_0(y)$ such that to a pair of points $(x, 0)$ and $(x^0, 0)$ there corresponds a finite number of curves. It should be noted that the phenomenon of ray focussing occurs for example, in seismological investigations of the Earth. By use of seismograms, geophysicists can already distinguish in some cases the signals coming to a point by different paths. Therefore the inverse linearized problem is meaningful also when there is no one-to-one correspondence between the parameters ξ, η and x, x^0.

IV. Consider now the case where $\mathscr{D}$ is a spherical layer $0<r_0\leqslant r\leqslant 1$ $r=|x|$, in $x_1, \ldots, x_m$ space and the function n_0 depends only on the distance to the origin: $n_0=n_0(r)$. Let the function $\tau_1(x. x^0)$ be specified at the points of the unit sphere $r=1$. In this case the geodesics in the metric $n_0(r)$ also represent plane curves lying in two-dimensional planes passing through the origin. Therefore, analogously to the foregoing, one can solve a plane problem in each cross-section of $\mathscr{D}$ in such planes. We introduce in this cross-section polar coordinates r, φ. Then, if the function $m_0(r)=rn_0(r)$ satisfies the inequality $m_0'(r)>0$, $r_0\leqslant r\leqslant 1$, to each point (ϱ, α) of the annulus $\mathscr{D}_1\{r_0\leqslant r\leqslant 1, 0\leqslant\varphi\leqslant 2\pi\}$ there corresponds one geodesic $\Gamma_0(\varrho, \alpha)$ for which the point (ϱ, α) is a vertex and which joins a pair of points of the circumference $r=1$. The equation of the geodesic $\Gamma_0(\varrho, \alpha)$ can be written in the form

$$\varphi=\alpha\pm\int_{\varrho}^{r}\frac{m_0(r)\,dr}{\sqrt{m_0^2(r)-m_0^2(\varrho)}}, \qquad (r_0\leqslant\varrho\leqslant r\leqslant 1). \tag{18}$$

Denote by $\tau_1(\varrho, \alpha)$ the value of the function τ_1 corresponding to the geodesic $\Gamma_0(\varrho, \alpha)$. Then, analogously to Sec. 6 of Chapter 1, for each Fourier harmonic $n_{1k}(r)$ of the function $n_1(r, \varphi)$ we construct equations of form (6.17):

$$\begin{gathered} n_{1k}(t)+\int_{t}^{1} n_{1k}(r)\,R_k(r, t)\,dr= f_k(t), \\ (r_0\leqslant t\leqslant 1)\; k=0, \pm 1, \pm 2, \ldots, \end{gathered} \tag{19}$$

in which the kernel $R_k(r, t)$ and right-hand sides $f_k(t)$ are given by $R_k(r, t)=$

$$-\frac{\sqrt{2m_0(t)}}{\pi\sqrt{m_0'(t)}}\frac{\partial}{\partial t}\int_{t}^{r}\frac{m_0(r)}{\sqrt{(\varrho-t)\,[m_0^2(r)-m_0^2(\varrho)]}}\cos k\int_{\varrho}^{r}\frac{m_0(\varrho)\,dr}{\sqrt{m_0^2(r)-m_0^2(\varrho)}}\,d\varrho,$$

$$f_k(t) = -\frac{\sqrt{m_0(t)}}{\pi\sqrt{2m_0'(t)}}\frac{\partial}{\partial t}\int_t^1 \frac{\tau_{1k}(\varrho)}{\sqrt{\varrho - t}}\,d\varrho. \tag{20}$$

$k = 0, \pm 1, \pm 2, \ldots$

Thus the following theorem is valid.

Theorem 2.11. *Let the function $n(x) \in C^2$ and be representable in the region $\mathcal{D}\{0 < r_0 \leqslant r \leqslant 1\}$, $r = |x|$, by*

$$n(x) = n_0(r) + n_1(x),$$

where $n_0(r) \in C^2$ is a known function such that $(rn_0(r))' > 0$, $r_0 \leqslant r \leqslant 1$, and the function $n_1(x)$ is small in the norm of C^2. Then the function $n_1(x)$ is uniquely determined in the region $\mathcal{D}$ and in terms of the function $\tau_1(x, x^0)$ whose arguments x and x^0 vary over the set $\mathfrak{M}\{r = 1\}$.

6. One-Dimensional Inverse Kinematic Problem for the Wave Equation in Anisotropic Media

I. Consider in $x_1, \ldots, x_m$ space the equation

$$u_{tt} = \sum_{i,j=1}^{m} a_{ij}(x)\, u_{x_i x_j} + \delta(x - x^0, t) \tag{1}$$

with the initial condition

$$u|_{t<0} \equiv 0. \tag{2}$$

It is of interest to find out what the formulation of an inverse kinematic problem can give for this equation. It was noted above that disturbances from a point source propagate along the geodesics in the metric

$$ds = \sqrt{\sum_{i,j=1}^{m} b_{ij}(x)\, dx_i\, dx_j}, \qquad \|b_{ij}\| = \|a_{ij}\|^{-1}. \tag{3}$$

The travel time of a disturbance from x^0 to x is determined by the formula

$$\tau(x, x^0) = \int_{\Gamma(x, x^0)} \sqrt{\sum_{i,j=1}^{m} b_{ij}\, dx_i\, dx_j}. \tag{4}$$

The derivative of the function $\tau(x, x^0)$ computed at the point x in the direction l has the obvious physical meaning of being a quantity inverse to the speed of propagation of a disturbance in the direction of l. At the same time

$$\frac{d\tau}{dl}=\sqrt{\sum_{i,j=1}^{m} b_{ij}\frac{dx_i}{dl}\frac{dx_j}{dl}} \tag{5}$$

and hence, the propagation speed of a disturbance at a given point depends on the direction. Therefore (1) describes the wave process in anisotropic media.

Let us consider a one-dimensional version of the inverse kinematic problem. Namely, let the coefficients a_{ij} of (1) be dependent only on one coordinate $x_m=y$ and let the function $\tau(x, x^0)$ be known for points belonging to the hyperplane $y=0$. Let $\mathscr{D}$ be a region of x, y space $(x=(x_1,\ldots, x_n))$ enclosed in $0\leqslant y\leqslant H$. It is required to find what information about the coefficients of (1) can be obtained in this case. It will be assumed here that the point source is fixed and is at the origin. In studying this problem we shall use the notation of Sec. 8 of Chapter 1.

It is quite clear that the above functionals can provide information about the coefficients of (1) in the region $\mathscr{D}$ if only the geodesics of metric (3) emanating from the origin, come within the region $\mathscr{D}$ and then return to the hyperplane $y=0$. Earlier we found a necessary and sufficient condition for such a behaviour of the geodesics. This condition amounts to the function $v'_\alpha(y)$ being positive in the interval $[0, H]$ for any unit vector $\alpha=(\alpha_1,\ldots, \alpha_n)$. Therefore the coefficients of (1) will be assumed to satisfy this condition. The physical meaning of this condition is that the propagation speed of a disturbance for each fixed direction parallel to the hyperplane $y=0$ should monotonically increase with increasing y. Indeed, let $l=(\alpha, 0)$ be a unit vector of x, y space. Then formula (5) can be rewritten as

$$\frac{d\tau}{dl}=\sqrt{\alpha\, B(y)\, \alpha^*}=\sqrt{\gamma_\alpha(y)}. \tag{5'}$$

Consider now two families of ellipsoids

$$z\, B^{-1}(y)\, z^*=1,$$
$$z\, B(y)\, z^*=1$$

dependent on parameter y. Take two arbitrary values y_1, $y_2\in[0, H]$. Let $y_1>y_2$; then, by the condition $v'_\alpha(y)>0$, the ellipsoid $z\, B^{-1}(y_1)\, z^*$ lies strictly inside $z\, B^{-1}(y_2)\, z^*$. But in this case, obviously, the ellipsoid $z\, B(y_2)\, z^*$ lies strictly inside $z\, B(y_1)\, z^*$, hence, $\gamma_\alpha(y_1)<\gamma_\alpha(y_2)$. Arguing more rigorously, we can easily show that in fact $\gamma'_\alpha(y)<0$. From formula (5′) it follows that the speed in the direction $l=(\alpha, 0)$ is monotonic increasing in y.

Denote by $\tau(x, y)$ the propagation time from the origin to the point (x, y) and consider the expression $\operatorname{grad}_x \tau(x, y)$. From variational calculus it is known [71] that at the points of an extremal $\Gamma(x, y)$,

$$\operatorname{grad}_x \tau(x, y) = \operatorname{grad}_{x_y} \Delta .$$

Since the last expression, as follows from formula (8.10) of Chapter 1, is constant and equals z,

$$\operatorname{grad}_x \tau(x, y) = z \tag{6}$$

is constant along the ray $\Gamma(x, y)$ joining the origin to the point (x, y). At the same time, by the hypothesis of the problem, the function $\tau(x, y)$ is known at the points of the hyperplane $y=0$. Therefore the constant z corresponding to the ray $\Gamma(x, 0)$ is easily found to be

$$z = \operatorname{grad}_x \tau(x, 0) . \tag{7}$$

Formula (7) defines a correspondence between the set of points of the hyperplane $y=0$ and the set Z of those z which correspond to the geodesics $\Gamma(x, 0)$ joining these points with the origin. Note that we are justified in writing formula (7) when the function $\tau(x, 0)$ has continuous partial derivatives. One need not require then that the function $\tau(x, 0)$ be single-valued; it can have, for example, cusps. Moreover, the condition of continuous differentiability of the function $\tau(x, 0)$ eliminates the case of intense focussing of the rays. At the end of the section we shall present a sufficient condition on the coefficients of (1) which will imply that to each pair of points of the hyperplane $y=0$ there will correspond only one ray.

In Sec. 8 of Chapter 1 it is stated that the set Z corresponding to the rays passing in the region $\mathscr{D}\{0 \leqslant y \leqslant H\}$ is contained between the ellipsoids

$$z\, B^{-1}(0)\, z^* = 1 , \qquad z B^{-1}(H)\, z^* = 1 .$$

Consider the segment of the half-line $z = \alpha t$, $|\alpha| = 1\ (t > 0)$, belonging to the set Z and the set of those points of the hyperplane $y=0$ which correspond to this segment under correspondence (7). The image of the segment $z = \alpha t$ from the set Z is a curve of the hyperplane $y=0$ going through the origin. Its parametric equation $x = x_\alpha(t)$ can be easily written starting from formula (8.13′) of Chapter 1,

$$x_\alpha(t) = 2 \int_0^\eta \frac{t\alpha\, B^{-1}(y)}{\sqrt{s(y)\,[1 - t^2 v_\alpha(y)]}}\, dy , \tag{8}$$

$$v_\alpha^{-1/2}(H) \leqslant t \leqslant v_\alpha^{-1/2}(0) ,$$

where the parameter η is related to t by

$$v_\alpha(\eta)=\frac{1}{t^2}. \tag{9}$$

The curve in the hyperplane $y=0$ determined by formula (8) will be termed the "curve of observation". To each point of this curve (to be more exact, to each value of the parameter t) there corresponds a ray joining it with the origin and passing inside the region $\mathscr{D}$. Denote by $\tau_\alpha(t)$ the propagation time along the ray with the parameter t. From formula (4) and formula (8.11′) of Chapter 1 we find

$$\tau_\alpha(t)=2\int_0^\eta \frac{dy}{\sqrt{s(y)\,[1-t^2 v_\alpha(y)]}}. \tag{10}$$

Introduce instead of the variable y the new variable

$$\zeta=\int_0^y \frac{dy}{\sqrt{s(y)}}. \tag{11}$$

Formula (11) establishes a one-to-one differentiable correspondence $y=y(\zeta)$, $\zeta=\zeta(y)$ between the variables y and ζ. Denote

$$\tilde{v}_\alpha(\zeta)=v_\alpha(y(\zeta)); \quad \xi=\zeta(\eta),$$

and make a change of variables in the integral of formula (10). Then formula (10) becomes

$$\tau_\alpha(t)=2\int_0^\xi \frac{d\zeta}{\sqrt{1-t^2\tilde{v}_\alpha(\zeta)}}. \tag{10′}$$

Moreover, it is obvious that

$$\tilde{v}_\alpha(\xi)=\frac{1}{t^2}. \tag{12}$$

From formula (12) it follows that there exists one-to-one correspondence between ξ and t.

Indeed,

$$\frac{d}{d\xi}\tilde{v}_\alpha(\xi)=[v'_\alpha(\eta)\ \sqrt{s(\eta)}]_{\eta=\eta(\xi)}>0,$$

and hence, the function $\tilde{v}_\alpha(\xi)$ increases monotonically. But then to each value of t there corresponds by formula (12) only one value $\xi = \mu_\alpha(t)$. Let us show now that the given information enables us to find this correspondence. Indeed, let $t_0(\alpha)$ be the value of the parameter t corresponding to the ray tangent to the hyperplane $y=0$, namely

$$t_0(\alpha) = v_\alpha^{-1/2}(0) = \tilde{v}_\alpha^{-1/2}(0) = \lim_{x_\alpha(t)\to 0} |\mathrm{grad}_x \tau(x, 0)|.$$

Apply to equality (10′) the operator

$$\mathrm{L}_\alpha \tau_\alpha = \frac{T}{\pi} \int_T^{t_0(\alpha)} \frac{\tau_\alpha(t)\, dt}{t\sqrt{t^2 - T^2}} \tag{13}$$

and make use of the following identity implied by formula (12):

$$\tilde{v}_\alpha(\mu_\alpha(t)) \equiv \frac{1}{t^2}.$$

Changing the order of integration in the resulting iterated integral we find

$$\begin{aligned}
\mathrm{L}_\alpha \tau_\alpha &= \frac{2T}{\pi} \int_T^{t_0(\alpha)} \frac{dt}{t\sqrt{t^2 - T^2}} \int_0^{\xi(t)} \frac{d\zeta}{\sqrt{1 - t^2 \tilde{v}_\alpha(\zeta)}} \\
&= \frac{2T}{\pi} \int_T^{t_0(\alpha)} \frac{dt}{t\sqrt{t^2 - T^2}} \int_{t_0(\alpha)}^{t} \frac{\mu_\alpha'(u)\, du}{\sqrt{1 - \dfrac{t^2}{u^2}}} \\
&= -\frac{2}{\pi} \int_T^{t_0(\alpha)} \mu_\alpha'(u)\, du \int_u^T \frac{Tu\, dt}{t\sqrt{(u^2 - t^2)(t^2 - T^2)}}.
\end{aligned}$$

The last integral is easily evaluated to obtain

$$\int_u^T \frac{Tu\, dt}{t\sqrt{(u^2 - t^2)(t^2 - T^2)}} = -\frac{\pi}{2}.$$

Finally taking into account that $\mu_\alpha(t_0(\alpha)) = 0$ we have the correspondence between the parameters ξ and t,

$$\mu_\alpha(T) = \mathrm{L}_\alpha \tau_\alpha. \tag{14}$$

But, since the correspondence between ξ and t, as pointed out above, is a one-to-one correspondence we hence know $t=t_\alpha(\xi)$. Thus formula (12) enables us to find the function $\tilde{v}_\alpha(\xi)$, namely,

$$\tilde{v}_\alpha(\xi)=\frac{1}{t_\alpha^2(\xi)}. \tag{15}$$

Thus, given the information about the propagation times on one "observation curve", we can find the function $\tilde{v}_\alpha(\xi)$. Consider $N=n(n+1)/2$ such "observation curves" corresponding to the unit vectors $\alpha_1, \ldots, \alpha_N$. Then we can find

$$\tilde{v}_{\alpha_k}(\xi)=\frac{1}{t_{\alpha_k}^2(\xi)}, \quad k=1, 2, \ldots, N. \tag{16}$$

Note that there are as many equations here as there are elements of matrix $\tilde{B}^{-1}(\zeta)=B^{-1}(y(\zeta))$. Considering the system of equations (16) as one for the elements of matrix $\tilde{B}^{-1}(\xi)$, we arrive for fixed ξ at the following geometric problem: given $N=n(n+1)/2$ points $z_k=t_{\alpha_k}(\xi)$, $k=1, 2, \ldots, N$, draw an ellipsoid through them with center at the origin:

$$z_k\tilde{B}^{-1}(\xi)\, z_k^*=1, \quad k=1, 2, \ldots, N. \tag{17}$$

This problem has a non-unique solution only in degenerate cases. Therefore, choosing the system of vectors $\alpha_1, \ldots, \alpha_N$, i.e. the system of appropriate "observation curves", we can uniquely determine matrix $\tilde{B}^{-1}(\xi)$. And this exhausts all of the information that the inverse kinematic problem can give about the coefficients of equation (1).

If the coefficient $s(y)$ is known, we can find $B^{-1}(y)$. If it is not known, the elements of matrix $B^{-1}(y)$ can be found within transformations in y corresponding to formula (11). When the coefficients $q(y)$ and $s(y)$ are expressed in terms of matrix $B^{-1}(y)$, the inverse kinematic problem enables us to find all the coefficients of (1). Indeed, in this case, knowing $\tilde{B}^{-1}(\zeta)$, we find $\tilde{s}(\zeta)=s(y(\zeta))$. But then from formula (11) we obtain the relation between ζ and y

$$y=\int_0^\zeta \sqrt{\tilde{s}(\zeta)}\, d\zeta \tag{17'}$$

and hence, we have $B^{-1}(y)$ and then $s(y)$ and $q(y)$. By use of this, it is not difficult to obtain the result of Herglotz-Wiechert mentioned in the previous section and the result of [6].

Summarizing, we have the following theorem,

Theorem 2.12. *Let the coefficients of* (1) *be dependent only on the variable y and satisfy*

1) *for any unit vector* $\alpha=(\alpha_1,\dots,\alpha_n)$, *the function* $v_\alpha(y)=\alpha B^{-1}(y)\,\alpha^*$ *has a positive derivative on the segment* $[0, H]$,

2) *the vector* $\operatorname{grad}_x \tau(x, 0)$ *is a continuous function in the hyperplane* $y=0$.

Then in the region $\mathscr{D}\{0\leqslant y\leqslant H\}$ *the elements of matrix* $\tilde{B}^{-1}(\zeta)$, *where* ζ *is related to* y *by formula* (11), *are uniquely determined in terms of the function* $\tau(x, 0)$. *If the coefficients* $s(y)$ *and* $q(y)$ *are expressed in terms of the elements of matrix* $B^{-1}(y)$, *then all the coefficients of* (1) *are found to be functions of the variable* y.

Above we already mentioned the physical meaning of the first condition. The result obtained also has a physical interpretation. Indeed, knowing matrix $B^{-1}(\zeta)$ one can find $B(\zeta)$ and then, using (5), the propagation speed of a disturbance in any direction $l=(\alpha, 0)$ as a function of the parameter ζ. Hence, if the propagation speed of a disturbance in any direction increases monotonically with increasing y, then it can be found as a function of l and ζ from the kinematic data in the hyperplane $y=0$.

Remark. The above theorem can be strengthened if we require, instead of the positiveness of $v'_\alpha(y)$ for all vectors α, its positiveness only for $N=n(n+1)/2$ vectors α_k such that the geometric problem about the drawing of the ellipsoid $z\,B^{-1}z^*=1$ through the points $z_k=\alpha_k t_k$ $(t_k>0)$, $k=1, 2,\dots, N$, corresponding to the vectors $\alpha_1,\dots,\alpha_N$, has a unique solution. It is true, though we cannot give a constructive solution of the problem because the "observation curves" are not known beforehand.

II. Consider now the following modified version of the inverse kinematic problem for (1) which also has practical application. Let the coefficients of (1) be functions of one variable y as before and twice continuously differentiable everywhere except at $y=H$ where they have discontinuities of the first kind. Then, if there is a point source at the origin, there will be apart from ordinary rays joining the points of the hyperplane $y=0$ to the origin, rays reflected from the hyperplane $y=H$. The reflection of rays occurs according to the law asserting the preservation of the magnitude of z. The following problem can then be posed. At the points of the hyperplane $y=0$, the propagation times of a disturbance from the source to these points are known along the rays reflected from the hyperplane $y=H$. Find the coefficients of (1) in the region $0\leqslant y\leqslant H$. It will be assumed that the coefficients of (1) are such in the interval $[0, H]$ that for any unit vector α the function $v_\alpha(y)$ is monotone $(v'_\alpha(y)\neq 0)$ and bounded: $0<m\leqslant v_\alpha(y)\leqslant M$.

It will also be assumed, analogously to the foregoing, that $\operatorname{grad}_x \tau(x, 0)$ is a continuous function. Then by formula (7) we can find the z corresponding to each ray and consider in the hyperplane $y=0$ the "observation curve" corresponding to the segment of the straight line $z=\alpha t$ of the set Z. The equation of this curve can be written in the form

$$x_\alpha(t)=2\int_0^H \frac{t\,\alpha\,B^{-1}(y)}{\sqrt{s(y)\,[1-t^2 v_\alpha(y)]}}\,dy. \tag{18}$$

The propagation time along the reflected ray joining the origin to $(x_\alpha(t), 0)$ is given by

$$\tau_\alpha(t)=2\int_0^H \frac{dy}{\sqrt{s(y)\,[1-t^2 v_\alpha(y)]}}. \tag{19}$$

Note that in this case the radicand in the denominator of the integral at least for a sufficiently small t (which corresponds to rays from the origin lying within a small angle from the y axis) is positive. We make the change of variables (11) in integral (19) and denote by ζ_0 the value of ζ corresponding to $y=H$. Then formula (19) becomes

$$\tau_\alpha(t)=\int_0^{\zeta_0} \frac{d\zeta}{\sqrt{1-t^2\tilde{v}_\alpha(\zeta)}}. \tag{20}$$

For small t, it is possible to expand the integrand in formula (20) in a uniformly convergent Taylor series in powers of $t^2 v_\alpha(\zeta)$. Carrying this out and changing the order of summation and integration, we find

$$\tau_\alpha(t)=\sum_{k=0}^{\infty} \frac{\frac{1}{2}(\frac{1}{2}+1)\dots(\frac{1}{2}+k-1)}{k!}\,t^{2k}\int_0^{\zeta_0} \tilde{v}_\alpha^k(\zeta)\,d\zeta.$$

From this equality it is easy to find

$$\int_0^{\zeta_0} \tilde{v}_\alpha^k(\zeta)\,d\zeta, \qquad (k=0,1,2,\dots), \tag{21}$$

which provide a complete system of moments for the function inverse to $\tilde{v}_\alpha(\zeta)$. The zero moment of this function is simply ζ_0. Referring to Hilbert's theorem from the theory of moments (see, for example, [2]), we can state that a monotonic function $\tilde{v}_\alpha(\zeta)$ is uniquely determined by the moments (21). Now it remains, as before, to consider $N=(n(n+1)/2$ suitable "observation curves" in order to find the matrix $\bar{B}^{-1}(\zeta)$. We determine, in addition, the parameter ζ_0. Thus in the present setting, a theorem analogous to Theorem 2.12 is valid. Here, it is true, it suffices to regard the function $v'_\alpha(y)$ as different from zero on the segment $[0, H]$.

III. We return now to the first formulation and find under what conditions on the coefficients of (1) there corresponds to each point on the "observation curve" exactly one ray joining it to the origin. Of course, the discussion will only be about sufficient conditions because the positiveness of the function $v'_\alpha(y)$ already guarantees the existence of a neighbourhood of the origin in which there is a uniqueness of correspondence between pairs of points and rays.

Consider an "observation curve" $x_\alpha(t)$. A sufficient condition for a one-to-one correspondence to exist between the parameter t and a point on this curve is that the projection of the vector $x_\alpha(t)$ onto the unit vector α increases monotonically with increasing values of t from $t_0(\alpha)=v_\alpha^{-1/2}(0)$ to $v_\alpha^{-1/2}(H)$. In other words, if the function

$$F(t)=x_\alpha(t)\ \alpha^*=2\int_0^{\eta(t)}\frac{tv_\alpha(y)\,dy}{\sqrt{s(y)\,[1-t^2v_\alpha(y)]}} \tag{22}$$

decreases monotonically with increasing t, then to each point on the curve $x_\alpha(t)$ there corresponds a unique value of t. It suffices now to find when the derivative of the function $F(t)$ is negative. For this purpose we make a change of variables in integral (22),

$$y=\mu_\alpha(u),$$

where $\mu_\alpha(u)$ is the function related to $v_\alpha(y)$ by

$$v_\alpha(\mu_\alpha(u))=\frac{1}{u^2}. \tag{23}$$

Integral (22) after this change becomes

$$F(t)=-2\int_{t_0(\alpha)}^{t}\frac{t\,\Phi(u)}{\sqrt{u^2-t^2}}\,du, \tag{24}$$

where

$$\Phi(u)=-\frac{\mu'_\alpha(u)}{u\sqrt{s(\mu_\alpha(u))}}. \tag{25}$$

The function $\Phi(u)$ is positive since by (23)

$$\mu'_\alpha(u) = -\frac{2}{u^3 v'_\alpha(\mu_\alpha(u))} < 0.$$

Let us show that a sufficient condition for the fulfillment of the inequality $F'(t) < 0$ is the following:

$$\frac{d}{du}[u\Phi(u)] = \Phi(u) + u\,\Phi'(u) \leqslant 0. \tag{26}$$

Indeed, making a change of variables in integral (24),

$$u = t\cosh\varphi,$$

we can reduce (24) to

$$F(t) = 2t \int_0^{\operatorname{arc\,cosh} t_0(\alpha)/t} \Phi(t\cosh\varphi)\,d\varphi.$$

Differentiating it with respect to t, we find

$$F'(t) = -\frac{2\Phi(t_0(\alpha))}{\sqrt{t_0^2(\alpha) - t^2}} + 2 \int_0^{\operatorname{arc\,cosh} t_0(\alpha)/t} [\Phi(u) + u\Phi'(u)]_{u = t\cosh\varphi}\,d\varphi.$$

From this equality and from the positiveness of $\Phi(u)$ it follows that $F'(t) < 0$ if (26) holds. Let us write condition (26) in a more suitable form by expressing it in terms of the functions $v_\alpha(y)$ and $s(y)$. Note that because of formula (23) the function $u\,\Phi(u)$ can be written as

$$u\Phi(u) = \left.\frac{2v_\alpha^{3/2}(y)}{v'_\alpha(y)\,s^{1/2}(y)}\right|_{y = \mu_\alpha(u)} \tag{27}$$

and therefore condition (26) is equivalent to

$$\frac{d}{dy}\frac{v'_\alpha(y)\,s^{1/2}(y)}{v_\alpha^{3/2}(y)} \geqslant 0. \tag{28}$$

Thus, the fulfilment of condition (28) is sufficient for a single geodesic to exist joining the origin with each point of the plane $y = 0$. When (1) coincides with (5.1) of the previous section, $s(y) = 1/n^2(y)$, $v_\alpha(y) = 1/n^2(y)$ and condition (28) reduces to the known condition $(\ln n(y))'' \geqslant 0$ already mentioned above.

7. Multidimensional Linearized Inverse Kinematic Problem for the Wave Equation in Anisotropic Media

We consider now for (6.1) a linearization of the inverse kinematic problem analogous to that of Sec. 5. In this case it will be assumed that the coefficients $a_{ij}(x)$, $x=(x_1,\dots,x_m)$, of (6.1) can be represented in the form

$$a_{ij}(x)=a^0_{ij}(x)+\lambda a^1_{ij}(x), \qquad (i,j=1,2,\dots,m) \tag{1}$$

where the coefficients $a^0_{ij}(x)$ are known and $a^1_{ij}(x)$ are small in the norm of space C^2, and λ is a perturbation parameter. We expand the function $\tau(x,x^0)$ in a series with respect to λ,

$$\tau(x,x^0)=\tau^0(x,x^0)+\lambda\tau^1(x,x^0)+\lambda^2\tau^2(x,x^0)+\cdots \tag{2}$$

We make use, now, of the equation for the function $\tau(x,x^0)$. In this case it has the form (see [71])

$$\sum_{i,j=1}^m a_{ij}(x)\,\tau_{x_i}\tau_{x_j}=1. \tag{3}$$

We substitute the expressions for the coefficients a_{ij} and the function $\tau(x,x^0)$ from formulas (1) and (2) into the above expression and equate like powers of λ. This yields for the functions $\tau^0(x,x^0)$ and $\tau^1(x,x^0)$ the equations

$$\sum_{i,j=1}^m a^0_{ij}(x)\,\tau^0_{x_i}\tau^0_{x_j}=1, \tag{4}$$

$$2\sum_{i,j=1}^m a^0_{ij}(x)\,\tau^0_{x_i}\tau^1_{x_j}+\sum_{i,j=1}^m a^1_{ij}(x)\,\tau^0_{x_i}\tau^0_{x_j}=0. \tag{5}$$

The first one shows that $\tau^0(x,x^0)$ is physically the travel time in the medium with the coefficients $a^0_{ij}(x)$ and therefore can be regarded as known. Geometrically, it is the distance between the points x and x^0 in the metric in which the element of arclength ds^0 is given by

$$ds^0=\sqrt{\sum_{i,j=1}^m b^0_{ij}(x)\,dx_i\,dx_j}, \tag{6}$$

$b^0_{ij}(x)$ being the elements of the matrix $\bar{B}^0$ inverse to $\bar{A}^0=\|a^0_{ij}\|$. From variational calculus it is known that along a geodesic $\Gamma_0(x,x^0)$ in the metric (6) the equalities

$$\frac{dx_j}{ds^0}=\sum_{i=1}^m a^0_{ij}(x)\,\tau^0_{x_i}, \qquad (j=1,2,\dots,m) \tag{7}$$

hold.

Taking into account the above equalities we can write formula (5) in the form

$$2\frac{d\tau^1}{ds^0}+\sum_{i,j=1}^{m} a_{ij}^1(x)\,\tau_{x_i}^0\tau_{x_j}^0=0. \tag{8}$$

Integrating the last equality along the geodesic $\Gamma_0(x, x^0)$ we obtain

$$\tau^1(x, x^0)=-\tfrac{1}{2}\int\limits_{\Gamma_0(x,x^0)}\sum_{i,j=1}^{m} a_{ij}^1(\xi)\,\tau_{\xi_i}^0\tau_{\xi_j}^0\,ds^0. \tag{9}$$

One can use formulas (7) to eliminate the partial derivatives of the function $\tau^0(x, x^0)$. From (7) we find

$$\tau_{x_i}^0=\sum_{k=1}^{m} b_{ik}^0(x)\frac{dx_k}{ds^0}. \tag{10}$$

Eliminating the function $\tau^0(x, x^0)$ from (9) we arrive at the following formula for $\tau^1(x, x^0)$:

$$\tau^1(x, x^0)=-\tfrac{1}{2}\int\limits_{\Gamma_0(x,x^0)}\sum_{k,l=1}^{m} c_{kl}(\xi)\frac{d\xi_k}{ds^0}\frac{d\xi_l}{ds^0}\,ds^0, \tag{11}$$

where

$$c_{kl}=\sum_{i,j=1}^{m} a_{ij}^1(x)\,b_{ik}^0(x)\,b_{jl}^0(x). \tag{12}$$

Formula (11) generalizes the corresponding formula in Sec. 5.

The linearized inverse kinematic problem, as we stated in Sec. 5, is as follows: The function $\tau^1(x, x^0)$ is regarded as known at the points x and x^0 of the surface S bounding the region $\mathscr{D}$; it is required to find the coefficients $a_{ij}^1(x)$ in terms of this function. From the previous section it follows that in this version it is, generally speaking, impossible to determine all the coefficients $a_{ij}^1(x)$, because in the case of a one-dimensional inverse problem it is impossible to find all the coefficients of the differential equation. This is explained by the fact that along the geodesic $\Gamma_0(x, x^0)$ the components of the tangent vector dx/ds^0 are not independent. They are related by the condition that the vector dx/ds^0 has unit length in the Riemannian metric:

$$\sum_{k,l=1}^{m} b_{kl}^0(x)\frac{dx_k}{ds^0}\frac{dx_l}{ds^0}=1. \tag{13}$$

Let us investigate the linearized inverse problem for the case when all coefficients a_{ij}^0 are dependent only on one variable. Let it be, as usual, the variable $x_m = y$. Consider in x, y space, $x = (x_1, \ldots, x_n)$, $n = m-1$, the region $\mathscr{D}\{0 \leqslant y \leqslant H\}$ and let the function $\tau^1(x, x^0)$ be known at the points of the hyperplane $y=0$. In the case of the one-dimensional formulations considered in the preceding section we saw already that the coefficients of (6.1) corresponding to the variable y cannot be found from kinematic information. Therefore assume that the coefficients a^1_{mj} $(j=1, 2, \ldots, m)$ corresponding to the variable y are known and that our task is to find the other coefficients $a^1_{ij}(x)$, $(i, j=1, 2, \ldots, n)$. Let the coefficients $a^0_{ij}(y)$ $(i, j=1, 2, \ldots, m)$ satisfy the conditions of Theorem 2.12. Then the curves $\Gamma_0(x, x^0)$ which in this case are known (their equations are given by formulas (24)–(26) of Sec. 8 in Chapter 1), behave regularly in the region $\mathscr{D}$ (in the sense of the unique solution of the integral-geometric problem on them). Consider in the hyperplane $y=0$ an "observation curve" going through the point $(x^0, 0)$ corresponding to the admissible values of $z = \operatorname{grad}_x \tau^0(x, 0)$ lying on the half-line $z = \alpha t (t > 0)$. The equation of this "observation curve" according to formula (6.8) has the form

$$x_\alpha(x^0, t) = x^0 + 2\int_0^\eta \frac{t\,\alpha\, B_0^{-1}(y)}{\sqrt{s_0(y)\,[1 - t^2 v_\alpha^0(y)]}}\, dy. \tag{14}$$

In this formula the notation of the previous section has been retained and the coefficients corresponding to metric (6) are furnished with a zero index. Introduce for the matrix $\bar{A}^1 = \|a^1_{ij}\|$, $i, j=1, 2, \ldots, m$, the notation

$$A^1 = \|a^1_{ij}\| \;(i, j=1, 2, \ldots, n); \qquad q^1 = \|a^1_{mj}\| \;(j=1, 2, \ldots, n); \qquad s_1 = a^1_{mm}.$$

Then the value of the function τ^1 corresponding to the ray $\Gamma_{0\alpha}(x^0, t)$ joining x^0 to $x_\alpha(x^0, t)$ of the "observation curve" can be written, according to formula (9), as

$$\tau^1_\alpha(x^0, t) = -\tfrac{1}{2}\int_{\Gamma_{0\alpha}(x^0, t)} z A^1 z^*\, ds^0 - \tfrac{1}{2}\int_{\Gamma_{0\alpha}(x^0, t)} [2q^1 z^* + s_1 \tau^0_y]\, \tau^0_y\, ds^0. \tag{15}$$

Note also that along the ray $\Gamma_{0\alpha}(x^0, t)$ the partial derivative τ^0_y can be expressed, according to formula (4), in terms of the components of $\operatorname{grad}_x \tau^0(x, y) = z$. The specific expression for τ^0_y is

$$\tau^0_y = -\frac{q^0\; z^*}{s_0} + \frac{1}{s_0 \Delta_0}, \tag{16}$$

where Δ_0 is given by formula (8.11′). On the strength of our assumption, the second integral in (15) can be regarded as known. Let x^0 vary over a point set of the plane $y=0$, choosing each time an "observation curve" corresponding to the fixed α. Since the coefficients a_{ij}^0, $i, j=1, 2, \ldots, m$, are dependent only on the coordinate y, the "observation curve" will be displaced parallel to itself as a rigid whole. The rays $\Gamma_{0\alpha}(x^0, t)$ are also invariant to motions of the point x^0. But then for the function $v_\alpha^1(x, y) = \alpha A^1 \alpha^*$ we arrive at the integral-geometric problem considered in Sec. 8 of Chap. 1. By Theorem 1.11 the function $v_\alpha^1(x, y)$ can be determined uniquely in the class of functions having compact support. If now we consider at each point x^0, $N=n(n+1)/2$ "observation curves" chosen so that the corresponding geometric problem discussed in Sec. 6 has a unique solution, then, having solved the integral-geometric problem N times, we find the N functions $v_{\alpha_1}^1, \ldots, v_{\alpha_N}^1$ from which we can then find the a_{ij}^1 $(i, j=1, 2, \ldots, n)$.

Thus we have established the following theorem.

Theorem 2.13. *Let the coefficients* a_{ij} *of* (6.1) *have the representation in the region* $\mathscr{D}\{0 \leqslant y \leqslant H\}$,

$$a_{ij}(x, y) = a_{ij}^0(y) + a_{ij}^1(x, y), \qquad (i, j=1, 2, \ldots, m)$$

where the functions $a_{ij}^0(y)$ *are known and satisfy the conditions of Theorem* 2.12 *and the functions* $a_{ij}^1(x, y)$ *have compact support in the region D and are small in the norm of space* C^2. *Then, given the functions* a_{mj}^1 $(j=1, 2, \ldots, m)$, *one can uniquely determine* a_{ij}^1 $(i, j=1, 2, \ldots, n)$ *in terms of the function* $\tau^1(x^0, 0, x, 0)$, *where* $(x^0, 0)$ *and* $(x, 0)$ *are arbitrary points of the hyperplane* $y=0$.

Chapter III

Application of the Linearized Inverse Kinematic Problem to Geophysics

This chapter presents some considerations concerning the possible application of the linearized inverse kinematic problem to the geophysical study of the inhomogeneous structure of the Earth. The numerical solution of the problem is described in a slightly simplified form and a number of numerical computer experiments performed with the aid of this technique are discussed.

1. The Earth's Structure from a Geophysical Standpoint and the Problem of Determining the Velocity Structure of the Earth's Mantle

The question of the internal structure of the Earth is of importance and interest from both a perceptual and a practical respect; it is directly related to the problem of finding deposits of minerals, the form of the Earth's topography, volcanic eruptions, earthquakes and so on. However a direct study of the Earth is only possible for a small portion of the globe up to depths of the order of 5—10 km. Therefore methods by which the deep layers of the Earth can be studied acquire great importance. Geophysicists have as of now several procedures available to them. Included among these are first of all the seismological, magnetic, gravimetric and electric methods. Of these the seismological method is the most economical and apparently most promising way of studying the Earth's core. It amounts to studying the elastic vibrations due to earthquakes or man-made explosions. Disturbances inside the Earth in the form of instantaneous ruptures cause elastic vibrations or elastic waves of two kinds (see [53]). Waves traveling with great velocities are the compression and rarefaction waves. They consist of oscillations which are in the direction of propagation of the wave and are called longitudinal waves. The second kind of waves are oscillations involving shift or torsion perpendicular to the direction of propagation of the waves and they are therefore called transverse waves. The propagation velocity of longitudinal and transverse waves are related to the elastic properties of the matter through which they are traveling by

$$v_p=\sqrt{\frac{\lambda+2\mu}{\varrho}}, \qquad v_s=\sqrt{\frac{\mu}{\varrho}}. \tag{1}$$

In these formulas, λ and μ are the Lamé constants characterizing the elastic properties of the medium; ϱ is the density; v_p and v_s are the propagation velocities of longitudinal and transverse waves, respectively.

An elastic medium is completely characterized by the parameters λ, μ and ϱ. Therefore if v_p and v_s were known, relations (1) would give a fairly large amount of information about the Earth's matter. In connection with this, one of the basic seismological problems is to determine the propagation velocities for longitudinal and transverse waves. As we have already stated before, one makes use of the observations obtained from oscillations in the Earth's surface due to earthquakes. The energy of some earthquakes is so great that their effect can be measured at any point of the globe. The process of elastic vibrations propagating in the Earth is described by the system of dynamical equations of elasticity theory. An analysis of them leads to the conclusion that longitudinal and transverse waves propagate inside the elastic Earth sphere according to the same laws by which light and sound waves travel, i.e., they are reflected and refracted at a boundary between media with different physical properties. The paths along which they propagate are the geodesics in the metric associated with the magnitude of the wave velocity. This implies in particular that the velocities v_p and v_s inside the Earth can be found by making use of the version of the inverse kinematic problem whose general formulation was given in Sec. 5 of Chapter 2. Observations taken at seismological stations make it possible to associate with each earthquake, surface travel-time curves for longitudinal and transverse waves (i.e., tables of the travel-times of seismic waves from the earthquake point to the station). It is known that seismic paths along which the earthquake disturbances propagate first travel through the deep layers of the Earth before they see daylight. Hence the travel-time curves do not bear integral information about the structure of the deep layers of the Earth. The greater the epicentral distance, the deeper generally speaking do the seismic paths penetrate inside the Earth. With detailed travel-time curves available from a large number of earthquakes, it is therefore possible to hope to obtain a picture of the distribution of the propagation velocities of seismic waves inside the Earth.

In 1905–1907, Herglotz and Wiechert first considered the inverse seismological kinematic problem. Assuming that the Earth is spherically symmetric and that the mechanical properties of the Earth's matter depend only on the distance r from the Earth's center, they worked out a method for obtaining the propagation velocities for seismic waves from the surface travel-time curve. The assumption about the spherical symmetry of the Earth was based on the observed geophysical fact that travel-time curves constructed for earthquakes occurring in various regions of the globe do not differ very much. Therefore to first approximation, it was natural to assume that different regions of the globe had the same general velocity structure not depending on geographic coordinates. This made it possible by using averaged travel-time curves to construct a picture of the velocity distribution of seismic waves propagating along a radius of the Earth. These first one-dimensional models of the velocity structure of the Earth were of great developmental value since they allowed one to draw conclusions about the Earth's structure. At the same time, they stimulated the formulation of new versions of inverse problems and in particular the use of the amplitudinal properties of seismic waves, travel-time curves for reflected waves etc. We shall not describe the procedures by which the velocities of waves traveling in the Earth are determined since this is not our problem. We shall merely give results of studying the internal structure of the Earth on the basis of seismological data.

The modern conception of the Earth is that it consists of three basic parts, a crust, a mantle and a core. The Earth's crust is a comparatively thin layer. Its average thickness is 30–50 km and is less beneath the ocean and greater under continents increasing especially in mountainous regions. The crust is considered to be most strongly distinguishable by its velocity properties. There is an essential difference in the propagation velocities of waves from one region to another. The crust is separated from the mantle by the Mohorovičić discontinuity along which the propagation velocity of a wave experiences a first-order discontinuity. The Earth's mantle is of the order of thickness of 3000 km and consists of solid amorphous matter. The Earth's core possesses certain rigidity properties, namely, no transverse waves pass through it. The velocity of transverse waves undergoes a jump in passing from the mantle to the core. It is natural to assume that the matter comprising the core differs from that of the mantle.

As of now, a number of geophysicists have constructed one-dimensional spherically symmetric models for the velocity distribution of seismic waves propagating in the Earth (see, for example, [12] and [33]). They were developed on the basis of diverse seismological data and various procedures were used in interpreting it. It is therefore natural that there should be some differences in these models. It is however interesting to note that all of these models differ on the average from one another on the order of magnitude of 10%, i.e., they are fairly close to one another. At the same time, there are currently geophysical facts which make it possible to assume the existence of inhomogeneities inside the Earth with respect to geographic coordinates at least up to depths of several hundred kilometers. Among these facts are systematic deviations of the travel-time curves constructed for various land masses from the averaged travel-time curves and the asymmetry of the gravitational and electromagnetic fields. Moreover, the deviations from the surface travel-time curves corresponding to spherically symmetric velocity distributions for elastic waves is fairly small. For example, for travel times of seismic waves from source to receiver of the order of 10–20 min., the deviations in the travel-time curves do not exceed 5–6 sec. This enables one to suppose that the velocity structure of the Earth also deviates little from the spherically symmetric model and that the deviations are of the order of magnitude of 10‰. Nevertheless, these small fluctuations in the velocity distribution of propagating seismic waves is of great interest to geophysicists since they can possibly assist in clarifying the mechanism by which the Earth's core evolved, the question of drifting continents and so on. Thus, the problem of determining the three-dimensional velocity structure of the Earth is one of the real problems of geophysics.

The most studied part of the globe up to now has been its crust. The methods created by G. A. Gamburcev, Ja. V. Riznicenko, I. S. Berzon and N. N. Puzyrev to interpret reflected and head waves make it possible to study the Earth's crust systematically to find mineral deposits. The first maps of the structure of the Earth's crust in the USSR have been carried out. The Earth's mantle has been studied considerably less. There have been just a few first attempts to give a qualitative appraisal of the inhomogeneity of the Earth's mantle with respect to geographic coordinates. However these attempts are still in their initial stages. In the meantime, there has been a sufficient amount of seismological material accumulated leading to a non-one-dimensional velocity structure for the Earth's mantle. To this end, one can use in particular the linearized inverse kinematic problem developed in Sec. 5 of Chap. II. We showed that if the function $n(r, \theta, \varphi)$ reciprocal to the propagation speed of a wave (longitudinal or transverse) can be represented by

$$n(r, \theta, \varphi) = n_0(r) + n_1(r, \theta, \varphi), \tag{2}$$

where $n_0(r)$ is known and $n_1(r, \theta, \varphi)$ is small as compared to $n_0(r)$, then the surface travel-time curve $\tau(\theta_0, \varphi_0, \theta, \varphi)$ from the source with coordinates $(1, \theta_0, \varphi_0)$ to points (θ, φ) on the unit sphere can be represented up to second order quantities by

$$\tau(\theta_0, \varphi_0, \theta, \varphi) = \tau_0(\theta_0, \varphi_0, \theta, \varphi) + \tau_1(\theta_0, \varphi_0, \theta, \varphi). \tag{3}$$

Here $\tau_0(\theta_0, \varphi_0, \theta, \varphi)$ is the travel-time curve for a medium with refraction index $n=n_0(r)$ and $\tau_1(\theta_0, \varphi_0, \theta, \varphi)$ is given by

$$\tau_1(\theta_0, \varphi_0, \theta, \varphi)= \int\limits_{\Gamma_0(\theta_0, \varphi_0, \theta, \varphi)} n_1(r, \theta, \varphi)\, dS_0, \tag{4}$$

and $\Gamma_0(\theta_0, \varphi_0, \theta, \varphi)$ is the ray joining the points (θ_0, φ_0) and (θ, φ) on the unit sphere in a medium with refraction index $n=n_0(r)$. If $n_0(r)$ is such that $(r\, n_0(r))'>0$, $r_0\leqslant r\leqslant 1$, then the function $n_1(r, \theta, \varphi)$ can be uniquely determined in the region $\mathscr{D}\{0\leqslant r_0\leqslant r\leqslant 1\}$ from the function $\tau_1(\theta_0, \varphi_0, \theta, \varphi)$. Thus knowing the travel-time curves for earthquakes distributed over the surface of a unit sphere, we can find the fluctuations in the velocity with respect to geographic coordinates inside any spherical layer $r_0\leqslant r\leqslant 1$. Observe that most of the current one-dimensional velocity models for the mantle satisfy the inequality $(r\, n_0(r))'>0$. True, there have recently appeared some rather valid doubts as to the fact that the velocity in the mantle increases monotonically with depth. Many geophysicists (E. F. Savarenskii, I. L. Nersesov, V. I. Keilis-Borok and others) have expressed their supposition of the existence of layers in the Earth's mantle having reduced velocity, or as they are usually called, waveguides. When the latter are present, the inverse kinematic problem with sources of disturbances distributed above waveguides does not exhibit single-valuedness [28]. Uniqueness can be preserved within the framework of kinematics only by distributing sources of disturbances in the layer of reduced velocity itself. For certain regions of the globe, the version of the inverse kinematic problem with sources distributed inside the mantle has practical importance. Thus, for example, in the zone through the Kurile Islands and Japan, earthquakes that have occurred over the past 100 years have filled out the whole volume stretching to a depth of 600 km. On the whole, earthquakes are primarily situated in the Earth's crust. Therefore the possible presence of waveguides of course essentially restricts the applicability of the kinematic approach to studying the Earth's mantle. Unfortunately, there are currently no other methods enabling one to study the mantle with such detailedness and without the inadequacies (in regards to nonuniqueness of the solution when waveguides are present) of the kinematic approach. In this connection we consider it feasible to apply the linearized version of the inverse kinematic problem for studying the mantle. Its applicability is completely justified at least in those regions of the globe where waveguides are absent or where there is a sufficiently large number of earthquakes situated inside the waveguides.

The linearized version of the inverse kinematic problem makes it possible in principle to develop a three-dimensional velocity model for the Earth. However at present it is practically impossible to develop a three-dimensional model due to the fact that earthquakes occur over the globe very irregularly. Most of them happen in three spherical zones: 1) Arabia-Pamir-Baikal-Kamchatka, 2) Kamchatka-Kuril Islands-Japanese Islands, 3) mountain chains of North and South America. Most seismological stations are concentrated in these zones. Therefore at present we can only hope to obtain a non-one-dimensional structure for the Earth's mantle in the three cross-sections of the globe by planes of a great circle. The latter is quite possible theoretically because the linearized three-dimensional inverse kinematic problem can be split into a number of two-dimensional problems. In what follows we describe one of the computational algorithms suitable for the practical solution of the problem.

2. Numerical Solution of the Linearized Inverse Kinematic Problem

In solving the inverse kinematic problem for the Earth's mantle, we shall assume that we can find the travel-time curve on the Earth's surface on a sphere of smaller radius removing the effect of the Earth's crust and topography. To first approximation it can be supposed that the crust is characterized by constant mean speed, and the recalculation can be easily performed. Let R be the radius of this sphere. Consider a cross-section of the sphere of radius R in a plane passing through its center and introduce in it the polar coordinates r, φ with the pole at the center of the sphere. The coordinate r will be measured in fractions of the radius. Let $\mathscr{D}$ be the layer $0<r_0\leqslant r\leqslant 1$. Let $n(r, \varphi)$ be expressible in the region $\mathscr{D}$ in the form (1.2) and let us suppose that we have information about the travel time of a disturbance from point sources lying on the circle $r=1$ and inside the region $\mathscr{D}$ to a series of points on the circumference. The latter henceforth will be termed receivers and their coordinates will be denoted by $(1, \varphi_k^0)$, $k=1, 2, \ldots, K$. The coordinates of sources will be denoted by (r_l^*, φ_l^*), $l=1, 2, \ldots, L$, and propagation time from the source with the coordinates (r^*, φ^*) to the receiver with the coordinates $(1, \varphi^0)$ will be denoted by $\tau(r^*, \varphi^*, \varphi^0)$. Let $\Gamma_0(r^*, \varphi^*, \varphi^0)$ be the ray joining the points (r^*, φ^*) and $(1, \varphi^0)$ in media with refraction index $n_0(r)$ and $\tau_0(r^*, \varphi^*, \varphi^0)$ the propagation time of the signal in this media from the source to the receiver. Assuming that $(r\, n_0(r))'>0$ we can write $\Gamma_0(r^*, \varphi^*, \varphi^0)$, the equation of the ray, as

$$\varphi=\alpha\pm\int_{\varrho}^{r}\frac{m_0(\varrho)\,dr}{r\sqrt{m_0^2(r)-m_0^2(\varrho)}}, \qquad m_0(r)=r n_0(r). \tag{1}$$

Here (ϱ, α) are the coordinates of the vertex of the ray $\Gamma(r^*, \varphi^*, \varphi^0)$ (i.e.

the point on the ray which is closest to the center of the circle). The parameters (ϱ, α) are related to the coordinates of the source and the receiver by

$$|\varphi^* - \varphi^0| = \int_{\varrho}^{1} \frac{m_0(\varrho)\, dr}{r\sqrt{m_0^2(r) - m_0^2(\varrho)}} \pm \int_{\varrho}^{r^*} \frac{m_0(\varrho)\, dr}{r\sqrt{m_0^2(r) - m_0^2(\varrho)}},$$

$$\varphi^* + \varphi^0 = 2\alpha - \int_{\varrho}^{1} \frac{m_0(\varrho)\, dr}{r\sqrt{m_0^2(r) - m_0^2(\varrho)}} \pm \int_{\varrho}^{r^*} \frac{m_0(\varrho)\, dr}{r\sqrt{m_0^2(r) - m_0^2(\varrho)}}. \tag{2}$$

The $(+)$ sign is to be taken if

$$|\varphi^* - \varphi^0| \geqslant \int_{r^*}^{1} \frac{m_0(r^*)\, dr}{r\sqrt{m_0^2(r) - m_0^2(r^*)}}, \tag{3}$$

otherwise we have to take the $(-)$ sign. The propagation time of the signal in the media with refraction index $n_0(r)$ is given by

$$\tau_0(r^*, \varphi^*, \varphi^0) = \int_{\varrho}^{1} \frac{m_0^2(r)\, dr}{r\sqrt{m_0^2(r) - m_0^2(\varrho)}} \pm \int_{\varrho}^{r^*} \frac{m_0^2(r)\, dr}{r\sqrt{m_0^2(r) - m_0^2(\varrho)}}. \tag{4}$$

The rule of signs here is the same.

Note that there is no one-to-one correspondence between the parameters (ϱ, α) and the coordinates of the source and receiver in arbitrary media. This was already stated in Sec. 5 of Chap. II. In particular, some one-dimensional velocity models of the Earth make it necessary to consider the situation where corresponding to fixed source and receiver there are three rays (and not one) and, hence, three different values of the parameters (ϱ, α). Thus ϱ is a three-valued function of $|\varphi^* - \varphi^0|$ composed of three single-valued branches.

Knowing the propagation times of disturbances between sources and receiver, we can easily compute the difference between the experimental and the propagation times determined by formula (4). This difference to times determined by formula (4) is easy to calculate. This difference to within higher order quantities than $n_1(r, \varphi)$ coincides with $\tau_1(r^*, \varphi^*, \varphi^0)$. Taking into account the relation between the functions n_1 and τ_1 determined above and calculating the element of arclength along the ray $\Gamma_0(r^*, \varphi^*, \varphi^0)$ by means of formula (1), we find

$$\frac{1}{R}\tau_1(r^*, \varphi^*, \varphi^0) = \int_{\varrho}^{1} n_1(r, \varphi) \frac{m_0(r)\, dr}{\sqrt{m_0^2(r) - m_0^2(\varrho)}} \pm \int_{\varrho}^{r^*} n_1(r, \varphi) \frac{m_0(r)\, dr}{\sqrt{m_0^2(r) - m_0^2(\varrho)}}, \tag{5}$$

and φ being given by formula (1). Formula (5) is the starting point for determining the function $n_1(r, \varphi)$. If the sources are located sufficiently close to the boundary $r=1$ of the region $\mathscr{D}$, we can simply put $r^*=1$ and apply the method of reducing (5) to a series of Volterra integral equations of the second kind for the Fourier harmonics of the function $n_1(r, \varphi)$. Moreover we must first approximate the function τ_1 in some way by passing from its discrete to a continuous representation. The latter involves certain inconveniences, and therefore in solving (5) we have used another algorithm.

Let the function $n_1(r, \varphi)$ be represented in the form of a finite Fourier series

$$n_1(r, \varphi) = \sum_{k=-M}^{M} n_k^1(r)\, e^{ik\varphi}, \tag{6}$$

where the Fourier coefficients $n_k^1(r)$ are polynomials of the degree no higher than N:

$$n_k^1(r) = \sum_{n=0}^{N} A_{kn}(1-r)^n, \qquad A_{(-k)n} = \bar{A}_{kn}, \qquad (k=0, \pm 1, \ldots, \pm M). \tag{7}$$

Here $\bar{A}_{kn}$ denotes the complex conjugate of A_{kn}. Substituting the expression for $n_1(r, \varphi)$ into (5) we find

$$\frac{1}{R}\tau_1(r^*, \varphi^*, \varphi^0) = \sum_{k=-M}^{M}\sum_{n=0}^{N} A_{kn}\, e^{ik\alpha}\psi_{kn}(\varrho, r^*), \tag{8}$$

where

$$\psi_{kn}(\varrho, r^*) = \int_{\varrho}^{1} (1-r)^n \exp\left\{-ik\int_{\varrho}^{r} \frac{m_0(\varrho)\, dt}{t\sqrt{m_0^2(t)-m_0^2(\varrho)}}\right\} \frac{m_0(r)}{\sqrt{m_0^2(r)-m_0^2(\varrho)}}\, dr$$

$$\pm \int_{\varrho}^{r^*} (1-r)^n \exp\left\{\pm k\int_{\varrho}^{r} \frac{m_0(\varrho)\, dt}{t\sqrt{m_0^2(t)-m_0^2(\varrho)}}\right\} \frac{m_0(r)}{\sqrt{m_0^2(r)-m_0^2(\varrho)}}\, dr. \tag{9}$$

Formula (8) provides a representation for the function τ_1 in terms of the system of functions $e^{ik\alpha}\psi_{kn}(\varrho, r^*)$. Knowing the function $\tau_1(r^*, \varphi^*, \varphi^0)$ at a sufficiently large number of points, we can find the coefficients A_{kn}, and, hence, the function $n_1(r, \varphi)$.

This consideration was used also in the algorithm of finding an approximate solution of (5). The functions $\psi_{kn}(\varrho, r^*)$ were computed and tabulated. The method of least squares was used for finding the coefficients A_{kn}. The possibility of using computational programs for more general regions than a concentric ring is indicated. For seismology the case where measurements are made only on an arc AB rather than on the entire circumference $r=1$ is more characteristic. It is natural to consider the problem of construction of the function $n_1(r, \varphi)$ only in the region $\mathscr{D}_0$ bounded by the circumferences $r=1$ and $r=r_0$ and laterally by the rays emanating from the points A, B (Fig. 1).

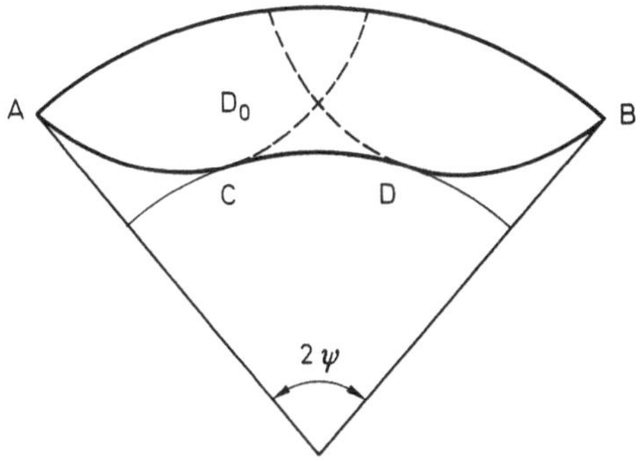

Fig. 1

Moreover, it is reasonable to take an expansion for $n_1(r, \varphi)$ in trigonometric functions of multiples of π/ψ. Here ψ is half the central angle measured by arc AB. When the coefficients A_{kn} are determined from equality (8), only the rays going through the region $\mathscr{D}_0$ need be taken into account.

3. Some Numerical Results

We shall outline here some results testing the above techniques through theoretical examples. For the model experiments we have taken the region $\mathscr{D}_0$ (Fig. 1) with the parameters $r_0=0.96$ and $\psi=\pi/10$. Such a choice of parameters is by no means accidental. It corresponds to an actual region of the globe lying in the cross-section of a great circle of the Earth. The arc AB of the region corresponds to the arc of the great circle going through Pamir and Lake Baikal. A large amount of seismological data has been accumulated for this region of the globe and it is proposed to use the data in the following to clarify the structure of the mantle in this region. The one-dimensional velocity distribution law with respect to depth is described here fairly well by the formula

$$n_0(r)=\frac{1}{28.07-r\cdot 20.33},$$

and it was used throughout the computations. Different sums of trigonometric functions of angular multiples of π/ψ with coefficients dependent on r are taken as the function $n_1(r, \varphi)$.

The sources φ_l^* and receivers φ_k^0 were placed at the points of arc AB at intervals of $\pi/100$. The distance between two neighbouring points on the Earth is then of the order of 200 km. The functions ψ_{kn} in this case ($r^*=1$) were computed from the formulas

$$\psi_{kn}(\varrho)=\int_{\varrho}^{1}(1-r)^n\frac{m_0(r)}{\sqrt{m_0^2(r)-m_0^2(\varrho)}}\cos 10k\int_{\varrho}^{r}\frac{m_0(\varrho)\,dt}{t\sqrt{m_0^2(t)-m_0^2(\varrho)}}\,dr,$$

$$(k=0, 1, \ldots, M; \quad n=0, 1, \ldots, N).$$

In Figs. 2–5 graphs of the functions $\bar{\psi}_{kn}(\varrho)=10^n\psi_{kn}(\varrho)$ for various k and n are given.

The experiments were conducted according to the following scheme. The function $n_1(r, \varphi)$ was specified in form (2.6). Then the propagation times $\tau(\varphi_l^*, \varphi_k^0)$ were found in terms of the function

$$n(r, \varphi)=n_0(r)+n_1(r, \varphi).$$

They were computed using A. V. Belonosova's program for solving the direct kinematic problem; the algorithm of the program is described in

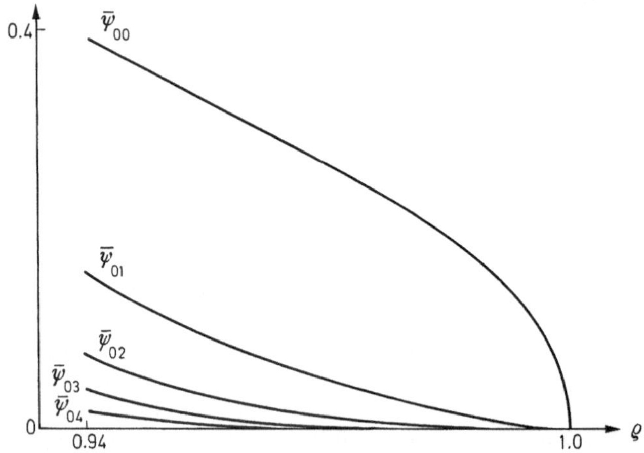

Fig. 2

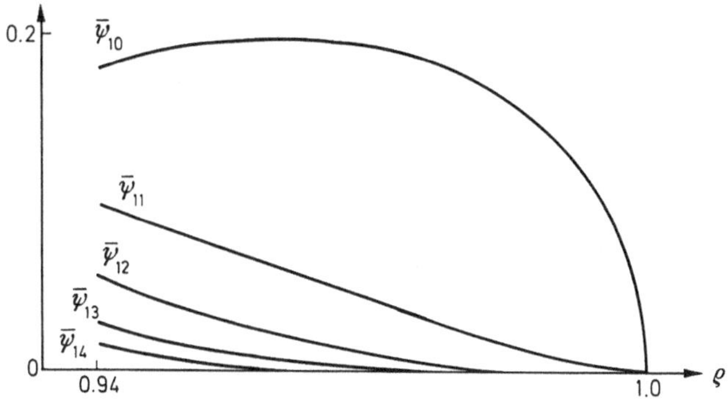

Fig. 3

Ref. [7]. The function $\tau_0(\varphi_l^*, \varphi_k^0)$ and then $\tau_1(\varphi_l^*, \varphi_k^0)$ were found by formula (2.4). After that the coefficients A_{kn} were found from formula (2.8) by the method of least squares and the function $n_1(r, \varphi)$ thus obtained was compared with the initial function.

Following are the results of calculation for four different media:

$$n_1(r, \varphi) = 0.004(\sin 10\varphi + r\cos 10\varphi + r\cos 20\varphi + \sin 30\varphi), \tag{1}$$

$$n_1(r, \varphi) = 0.0024 \sin 10\varphi \cdot \sin \frac{\pi(1-r)}{0.03}, \tag{2}$$

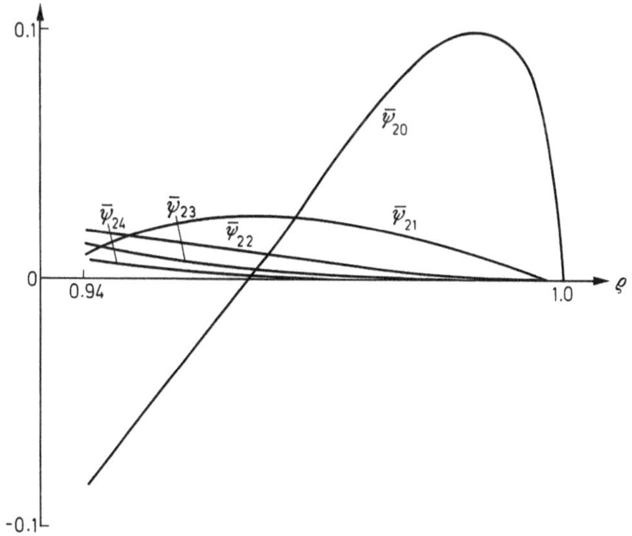

Fig. 4

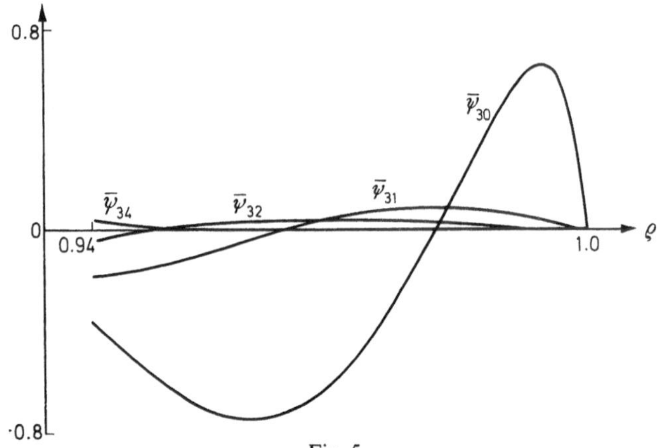

Fig. 5

$$n_1(r, \varphi) = 0.0009(\cos 10\varphi + \sin 20\varphi)\left[1 + \left(\frac{1-r}{0.06}\right)^2 + \sin\frac{\pi(1-r)}{0.03}\right], \quad (3)$$

$$n_1(r, \varphi) = 0.001 \sin 10\varphi \cdot \sin\frac{\pi(1-r)}{0.06}. \quad (4)$$

In all these, the function $n_1(r, \varphi)$ does not surpass 10% of the function $n_0(r)$. The No. of the type is in the upper right hand corner of each figure.

In Figs. 6–11 plots in different cross-sections $\varphi=$const. or $r=$const.

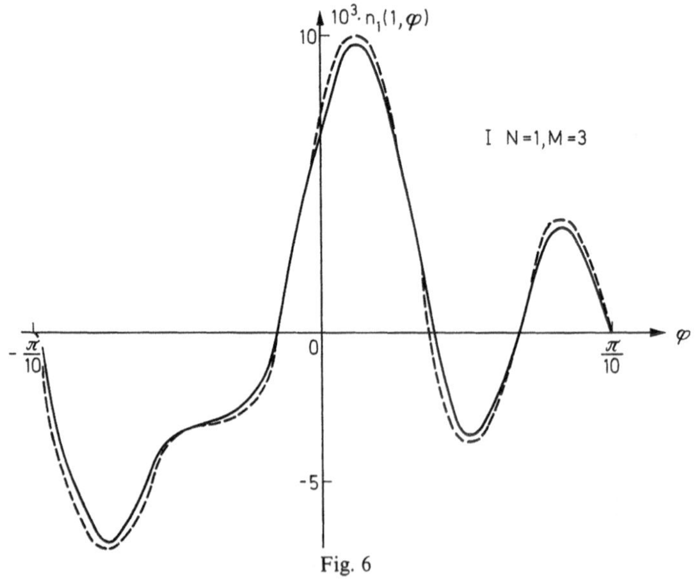

Fig. 6

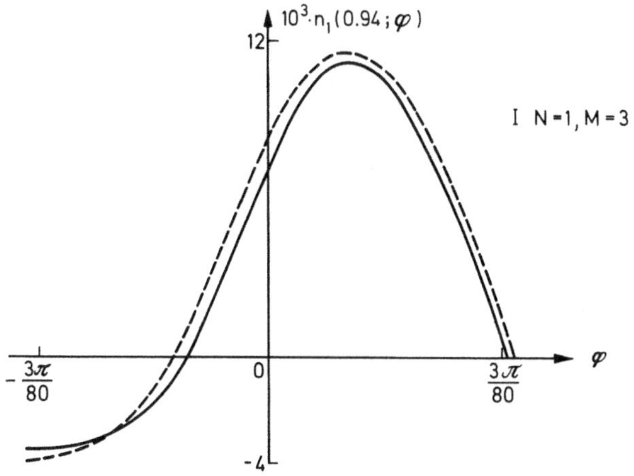
$10^3 \cdot n_1(0.94;\varphi)$
12
I N=1, M=3
$-\frac{3\pi}{80}$
0
$\frac{3\pi}{80}$
φ
-4

Fig. 7

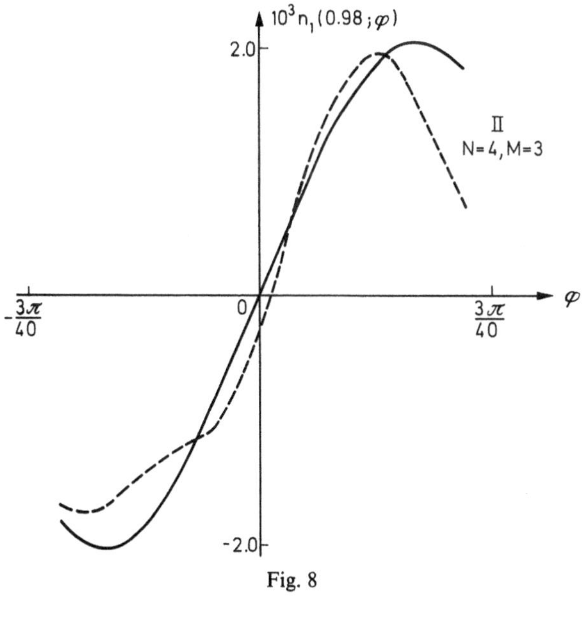
$10^3 n_1(0.98;\varphi)$
2.0
II
N=4, M=3
$-\frac{3\pi}{40}$
0
$\frac{3\pi}{40}$
φ
-2.0

Fig. 8

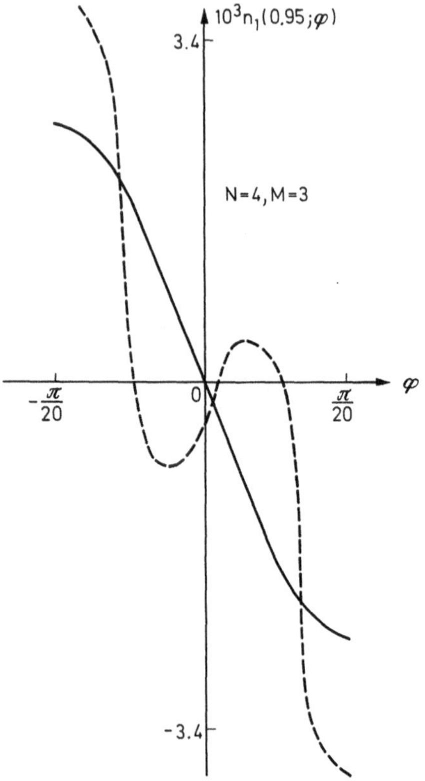

$10^3 n_1(0.95;\varphi)$
3.4
N=4, M=3
$-\frac{\pi}{20}$
0
$\frac{\pi}{20}$
φ
-3.4

Fig. 9

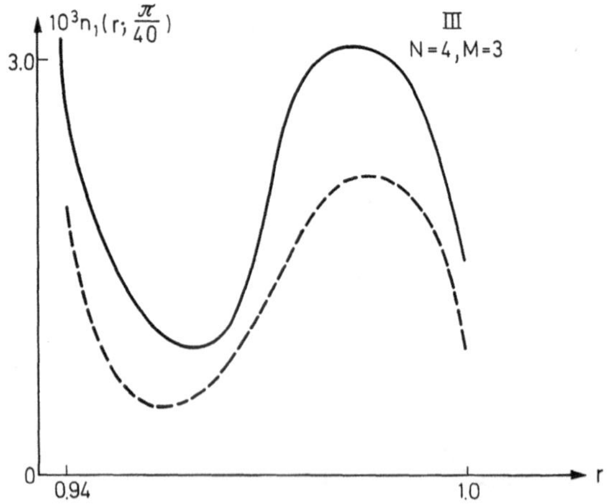

$10^3 n_1(r;\frac{\pi}{40})$
3.0
III
N=4, M=3
0
0.94
1.0
r

Fig. 10

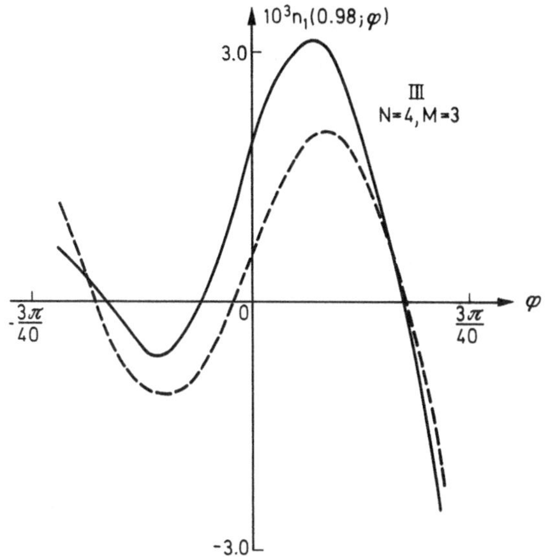

Fig. 11

of the region $\mathscr{D}$ are shown. The plot of the true function $n_1(r, \varphi)$ is represented by a solid line and that of the approximate function by a dotted line. The values of the parameters N, M on which the approximate solution is based are given. From these figures one can see that on the whole the approximate function qualitatively reflects the behaviour of the initial function and that the value of the function $n_1(r, \varphi)$ is fairly good in the middle of the region $\mathscr{D}_0$ and worse at points close to the boundary consisting of the arcs AC, BD, CD. This, however, is natural since the rays joining the sources and the receivers of the arc AB fill up most densely exactly the middle of $\mathscr{D}_0$.

A numerical study of the stability of the solution depending on the parameters N, M was made. The results are presented in Figs. 12–17. From these calculations one can see the following: if the number of the approximating functions is not enough to affect details of the behaviour of the function $n_1(r, \varphi)$, its determination is made on the average, and then with the increase of the number of the parameters a satisfactory approximation is reached which is maintained with further increase of N, M.

Stability of the solution of the problem under small variations in the initial data was also studied. For this purpose in the binary representation of the times $\tau(\varphi_1^*, \varphi_k^0)$, the binary digits 18, 17, ..., 12, 11 places after the decimal point were successively discarded, corresponding to errors 0.03, 0.05, 0.1, 0.2, 0.4, 0.08, 1.6, 3.2 sec. with reference to the Earth. The numerical analyses of the stability are shown in Figs. 18–21. The solution

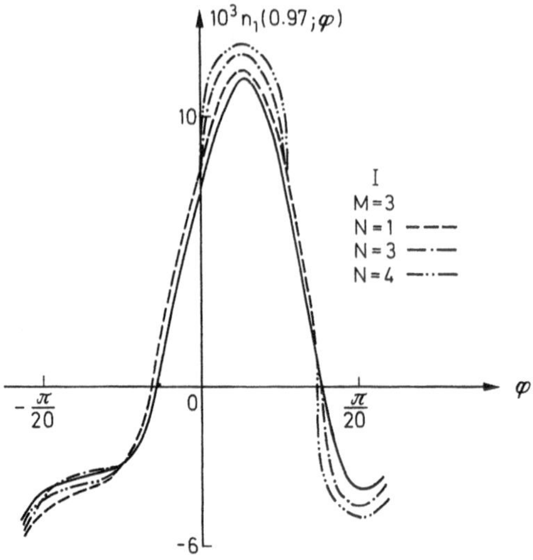

$10^3 n_1(0.97;\varphi)$
10
0
-6
$-\frac{\pi}{20}$
$\frac{\pi}{20}$
φ
I
M=3
N=1
N=3
N=4

Fig. 12

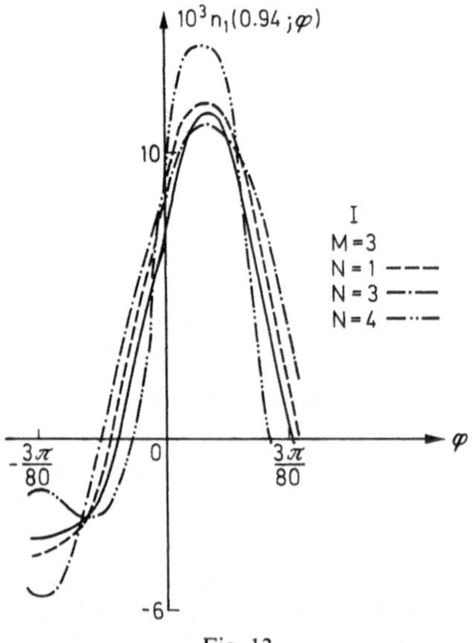

$10^3 n_1(0.94;\varphi)$
10
0
-6
$-\frac{3\pi}{80}$
$\frac{3\pi}{80}$
φ
I
M=3
N=1
N=3
N=4

Fig. 13

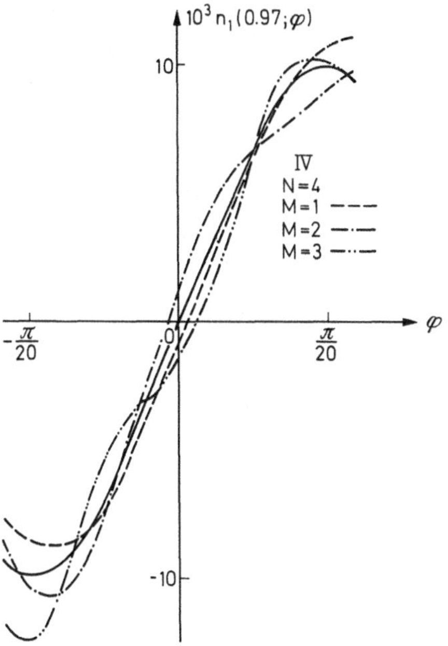
$10^3 n_1(0.97;\varphi)$
10
-10
IV
N=4
M=1
M=2
M=3
$-\frac{\pi}{20}$
0
$\frac{\pi}{20}$
φ

Fig. 14

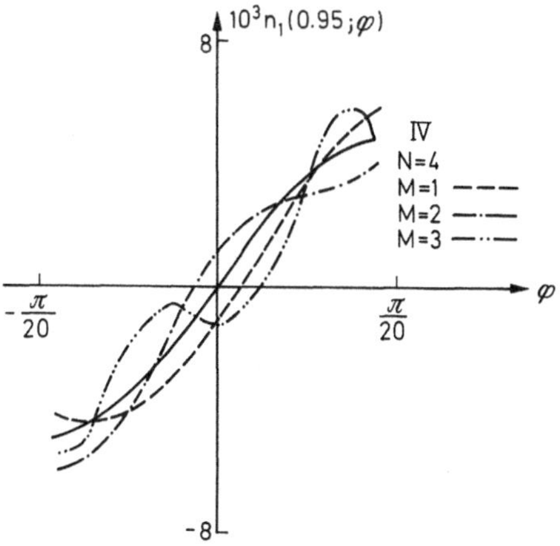
$10^3 n_1(0.95;\varphi)$
8
-8
IV
N=4
M=1
M=2
M=3
$-\frac{\pi}{20}$
$\frac{\pi}{20}$
φ

Fig. 15

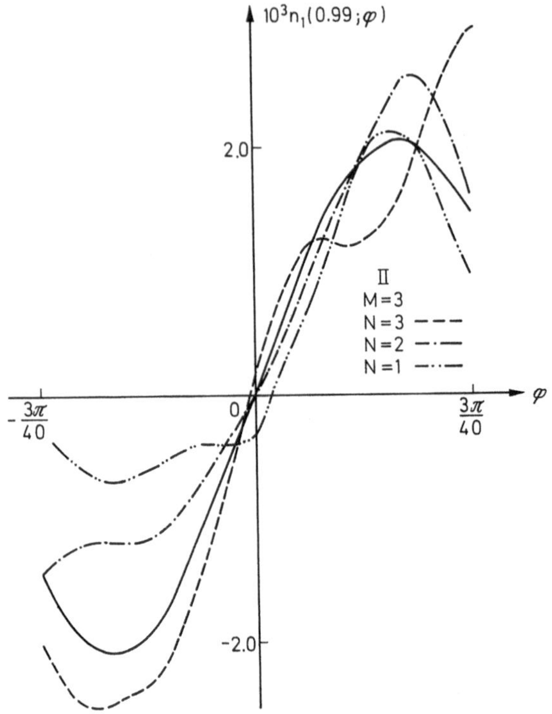

$10^3 n_1(0.99;\varphi)$
2.0
-2.0
II
M=3
N=3
N=2
N=1
$-\frac{3\pi}{40}$
0
$\frac{3\pi}{40}$
φ

Fig. 16

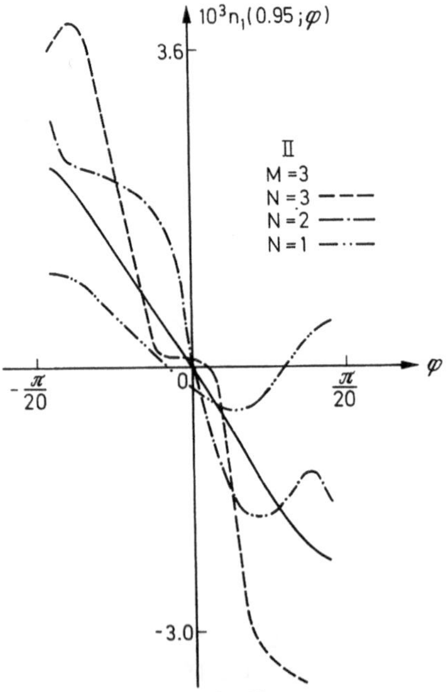

$10^3 n_1(0.95;\varphi)$
3.6
-3.0
II
M=3
N=3
N=2
N=1
$-\frac{\pi}{20}$
0
$\frac{\pi}{20}$
φ

Fig. 17

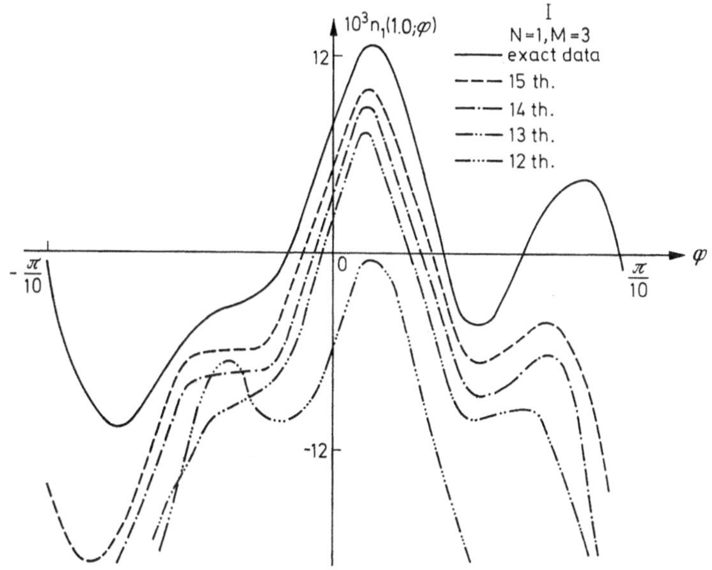
$10^3 n_1(1.0;\varphi)$
I
N=1, M=3
exact data
15 th.
14 th.
13 th.
12 th.
12
-12
0
$-\frac{\pi}{10}$
$\frac{\pi}{10}$
φ

Fig. 18

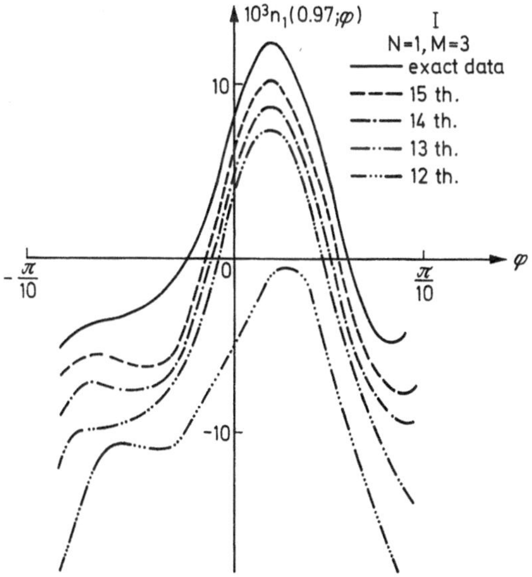
$10^3 n_1(0.97;\varphi)$
I
N=1, M=3
exact data
15 th.
14 th.
13 th.
12 th.
10
-10
0
$-\frac{\pi}{10}$
$\frac{\pi}{10}$
φ

Fig. 19

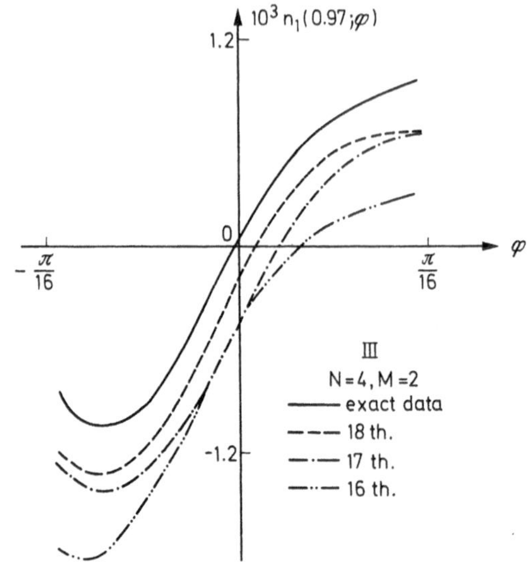
$10^3 n_1(0.97;\varphi)$
1.2
0
-1.2
$-\frac{\pi}{16}$
$\frac{\pi}{16}$
φ
III
N=4, M=2
exact data
18 th.
17 th.
16 th.

Fig. 20

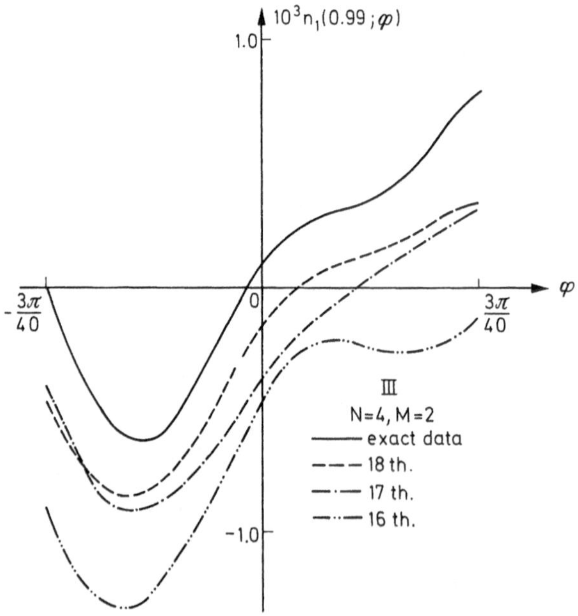
$10^3 n_1(0.99;\varphi)$
1.0
0
-1.0
$-\frac{3\pi}{40}$
$\frac{3\pi}{40}$
φ
III
N=4, M=2
exact data
18 th.
17 th.
16 th.

Fig. 21

corresponding to the exact initial data is represented by a solid line. As is obvious from the comparison of the results of calculation the characteristic behaviour of the solution in the case of type 1 media (Figs. 18–19) is maintained if errors do not exceed 0.8 sec. and in the case of type 4 if errors do not exceed 0.1 sec. Such a difference in admissible errors is simply explained by the fact the maximum value of $n_1(r, \varphi)$ for type 1 is ten times that for type 4. In both cases the admissible error is reached when in the normalized mantissa of the quantities $\tau_1(\varphi_i^*, \varphi_k^0)$ only 2–3 of the first binary places remain valid.

The numerical experiments performed using the theoretical results show that the above algorithm can be used for constructing inhomogeneous velocity models of the Earth's mantle. The inhomogeneities in the structure of the mantle stretching along geographical coordinates for several hundred kilometers can thus be taken into account.

Bibliography

(Starred items are in Russian)

1. Agranovič, Z. S., Marčenko, V. A.: The Inverse Problem of Scattering Theory. New York: Gordon and Breach Science Publishers 1963.
2.* Ahiezer, N. I.: The Classical Moment Problem. Moscow: Fizmatgiz 1961.
3.* Alekseev, A. S.: Some inverse problems in wave propagation theory. Izv. Akad. Nauk SSSR Ser. Geofiz. **11**, 1514–1531 (1962).
4.* Alekseev, A. S., Lavrent'ev, M. M., Muhometov, R. G., Nersesov, I. L., Romanov, V. G.: A numerical method for investigating horizontal inhomogeneities in the Earth's crust. Report to the Tenth General Assembly of the European Seismological Commission. Leningrad: September 1968.
5. Ambarcumjan, V. A.: Über eine Frage der Eigenwerttheorie. Z. Physik **53**, 690–695 (1929).
6.* Anikonov, Ju. E.: Some Problems in the Calculus of Variations and Integral Geometry. Novosibirsk: Computing Center, Sibirsk. Otdel. Akad. Nauk SSSR 1968, Preprint.
7.* Belonosova, A. V., Alekseev, A. S.: On a version of the inverse seismic kinematic problem for a twodimensional continuous inhomogeneous medium. In the collection: Some Methods and Algorithms for Interpreting Geophysical Data, pp. 137–154. Moscow: Nauka 1967.
8.* Belonosova, A. V., Tadžimuhamedova, S. S., Alekseev, A. S.: On the calculation of time-travel curves and of the geometric ray divergence in inhomogeneous media. In the collection: Some Methods and Algorithms for Interpreting Geophysical Data, pp. 124–136. Moscow: Nauka 1967.
9.* Berezanskiĭ, Ju. M.: A uniqueness theorem for the inverse spectral problem for Schrödinger's equation. Trudy Moskov. Mat. Obšč. **7**, 3–51 (1958).
10.* Blagoveščenskiĭ, A. S.: On the characteristic problem for an ultrahyperbolic equation. Mat. Sb. **63** (105), 1, 137–168 (1964).
11.* Blagoveščenskiĭ, A. S.: On the inverse problem of seismic wave propagation theory. In the collection: Problems in Mathematical Physics, No. 1. Leningrad: Izd-vo Leningrad. Univ. 1966.
12.* Bončkovskiĭ, V. F.: The Structure of the Earth's Interior. Moscow: Izd. Akad. Nauk SSSR 1953.
13. Borg, G.: Eine Umkehrung der Sturm-Liouville Eigenwertaufgabe; Bestimmung der Differentialgleichung durch die Eigenwerte. Acta Math. **78**, 2, 1–45 (1945).
14.* Budak, B. M., Iskanderov, A. D.: A difference method for solving coefficient boundary value problems. Dokl. Akad. Nauk SSSR **171**, 5, 1054–1057 (1966).
15.* Budak, B. M., Iskanderov, A. D.: On a class of boundary value problems with unknown coefficients. Dokl. Akad. Nauk SSSR **175**, 1, 13–16 (1967).

16.* Budak, B. M., Iskanderov, A. D.: On a class of inverse boundary value problems with unknown coefficients. Dokl. Akad. Nauk SSSR **176**, 1, 20–23 (1967).
17. Bullen, K. E.: Introduction to Theoretical Seismology, 3rd ed. Cambridge: University Press 1963.
18. Courant, R., Hilbert, D.: Methods of Mathematical Physics, Vol. II. New York: Interscience 1962.
19.* Čudov, L. A.: The inverse Sturm-Liouville problem. Mat. Sb. **25** (67), 451–456 (1949).
20.* Čudov, L. A.: A new version of the inverse Sturm-Liouville problem in a finite interval. Dokl. Akad. Nauk SSSR **109**, 1, 40–43 (1956).
21. Faddeev, L. D.: The inverse problem of quantum scattering theory. J. Mathematical Phys. **4**, 1, 72–104 (1963).
22.* Faddeev, L. D.: Properties of the S-matrix of the onedimensional Schrödinger equation. Trudy Mat. Inst. Steklov **73**, 314–335 (1964).
23.* Gamburcev, G. A.: Foundations of Seismological Prospecting. Moscow: Gostoptehizdat 1959.
24.* Gel'fand, I. M.: Integral geometry and its connection with representation theory. Uspehi Mat Nauk **15**, 2 (92), 155–160 (1960).
25. Gel'fand, I. M., Graev, M. I., Vilenkin, N. Ja.: Generalized Functions. Vol. 5: Integral Geometry and Representation Theory. New York: Academic Press 1965.
26.* Gel'fand, I. M., Levitan, B. M.: On the determination of a differential equation from its spectral function. Izv. Akad. Nauk SSSR Ser. Mat. **15**, 309–360 (1951).
27. Gel'fand, I. M., Šilov, G. E.: Generalized Functions. Vol. 1: Properties and Operations. New York: Academic Press 1964.
28.* Gerver, M. L., Markuševič, V. M.: Investigation into the lack of uniqueness in the determination of the velocity distribution of seismic waves from travel-time curves. Dokl. Akad. Nauk SSSR **163**, 6, 1337–1380 (1965).
29.* Gerver, M. L., Markuševič, V. M.: On the characteristic properties of seismic travel-time curves. Dokl. Akad. Nauk SSSR **175**, 2, 334–337 (1967).
30.* Gogoladze, V. G.: Cauchy's problem for the "generalized" wave equation. Dokl. Akad. Nauk SSSR **1**, 4, 166–169 (1934).
31.* Gogoladze, V. G.: The general problem of integrating the generalized wave equation with variable coefficients. Dokl. Akad. Nauk SSSR **1**, 6, 301–304 (1934).
32.* Gogoladze, V. G.: The wave equation for inhomogeneous and anisotropic media. Trudy Mat. Inst. Steklov **9**, 107–166 (1935).
33. Gutenberg, B.: Physics of the Earth's Interior. New York: Academic Press 1959.
34. Herglotz, G.: Über die Elastizität der Erde bei Berücksichtigung ihrer variablen Dichte. Z. für Math. Phys. **52**, 3, 275–299 (1905).
35.* Iskanderov, A. D.: On inverse boundary value problems with unknown coefficients for certain quasilinear equations. Dokl. Akad. Nauk SSSR **178**, 5, 999–10003 (1968).
36.* Ivanov, V. K.: The inverse potential problem for a body close to a given one. Izv. Akad. Nauk SSSR Ser. Mat. **20**, 793–818 (1956).
37.* Ivanov, V. K.: On the existence of a solution to the inverse logarithmic potential problem in finite form. Dokl. Akad. Nauk SSSR **106**, 598–599 (1956).
38.* Ivanov, V. K.: A uniqueness theorem in the inverse logarithmic potential problem for star-shaped sets. Izv. Vuzov Ser. Mat. **3**, 4 (1958).
39. John, F.: Plane Waves and Spherical Means Applied to Partial Differential Equations. New York: Interscience 1955.
40*. Kreĭn, M. G.: Solution of the inverse Sturm-Liouville problem. Dokl. Akad. Nauk SSSR **76**, 1, 21–24 (1951).
41.* Kreĭn, M. G.: Determination of the density of an inhomogeneous symmetric string from its frequency spectrum. Dokl. Akad. Nauk SSSR **76**, 3, 345–348 (1951).
42.* Kreĭn, M. G.: On inverse problems for an inhomogeneous string Dokl. Akad. Nauk

SSSR **82**, 5, 669–672 (1952).
43.* Kreĭn, M. G.: On the transition function for a onedimensional boundary value problem of second order. Dokl. Akad. Nauk SSSR **88**, 3, 405–408 (1953).
44.* Kreĭn, M. G.: An effective method for solving an inverse boundary value problem. Dokl. Akad. Nauk SSSR **94**, 6, 767–770 (1954).
45. Lavrent'ev, M. M.: A question concerning the inverse potential problem. Dokl. Akad. Nauk SSSR **106**, 1 (1956).
46. Lavrent'ev, M. M.: Some Improperly Posed Problems of Mathematical Physics, New York: Springer 1967.
47.* Lavrent'ev, M. M.: An inverse problem for the wave equation. Dokl. Akad. Nauk SSSR **157**, 3 (1964).
48.* Lavrent'ev, M. M.: On a class of inverse problems for differential equations. Dokl. Akad. Nauk SSSR **160**, 1, 32–35 (1965).
49.* Lavrent'ev, M. M, Romanov, V. G.: On three linearized inverse problems for hyperbolic equations. Dokl. Akad. Nauk SSSR **171**, 6, 1278–1281 (1966).
50. Lavrent'ev M. M., Romanov, V. G., Vasil'ev, V. G.: Multidimensional Inverse Problems for Differential Equations. New York: Springer 1970.
51.* Levitan, B. M.: On the asymptotic behavior of the spectral function and eigenfunction expansions for the equation $\Delta u + \{\lambda - q(x_1, x_2, x_3)\}\, u = 0$. Trudy Moskov. Mat. Obsc. **4**, 237–290 (1955).
52.* Levitan, B. M.: Generalized Shift Operators with Applications. Moscow: Fizmatigiz 1962.
53. Love, A. E. H.: A Treatise on the Mathematical Theory of Elasticity, 4th ed. London: Macmillan 1927.
54.* Marčenko, V. A.: Some theoretical questions concerning a second-order differential operator. Dokl. Akad. Nauk SSSR **72**, 3, 457–460 (1950).
55.* Marčenko, V. A.: Some theoretical questions concerning onedimensional second-order linear differential operators. Trudy Moskov. Mat. Obšč. **1**, 327–420 (1952), **2**, 3–82 (1953).
56.* Marčenko, V. A.: Stability in the inverse problem of scattering theory. Mat. Sb. **77** (119), 2, 139–162 (1968).
57.* Marčuk, G. I.: On the formulation of inverse problems. In the collection: Weather Prediction. Novosibirsk 1964.
58.* Mjunctc, G.: Integral Equations. Part 1. Moscow-Leningrad: GTTL 1934.
59.* Novikov, P. S.: On uniqueness for the inverse potential problem. Dokl. Akad. Nauk SSSR **18**, 165–168 (1938).
60.* Prilepko, A. I.: On the Theory of Inverse Problems for Geralized Potentials. Doctoral Dissertation. Novosibirsk: Mathematics Institute 1968.
61.* Rapoport, I. M.: On the twodimensional inverse problem in potential theory. Dokl. Akad. Nauk SSSR **28** (1940).
62.* Rapoport, I. M.: On stability in the inverse potential problem. Dokl. Akad. Nauk SSSR **31**, 4 (1941).
63.* Raševskiĭ, P. K.: Riemannian Geometry and Tensor Analysis. Moscow: Nauka 1967.
64.* Romanov, V. G.: On a linearized inverse dynamic problem for the telegraph equation. In the collection: Methods and Algorithms for Interpreting Geophysical Data. Moscow: Nauka 1967.
65.* Romanov, V. G.: On the determination of a function from its integrals over ellipsoids of revolution with one fixed focus. Dokl. Akad. Nauk SSSR **173**, 4, 766–769 (1967).
66.* Romanov, V. G.: On the determination of a function from its integrals over a family of curves. Sibirsk. Mat. Ž. **4**, 1, 1206–1208 (1968).
67.* Romanov, V. G.: A onedimensional inverse problem for the telegraph equation.

Differential Equations. **4**, 1, 87–101 (1968).

68.* Romanov, V. G.: Some inverse problems for hyperbolic differential equations. Report to the Sixth All-Union Acoustics Conference. Moscow 1968.

69.* Romanov, V. G.: On an inverse problem for the generalized wave equation. Dokl. Akad. Nauk SSSR **181**, 3, 554–557 (1968).

70.* Romanov, V. G.: An integral geometric problem and a linearized inverse problem for differential equations. Sibirsk. Mat. Ž. **10**, 6, 1364–1374 (1969).

71. Smirnov, V. I.: Course of Higher Mathematics. Vol. 4, Moscow: Gostehizdat 1956.

72.* Sobolev, S. L.: The wave equation for inhomogeneous media. Trudy Seismolog. Inst. **6**, 1–57 (1930).

73.* Sobolev, S. L.: On the integration of the wave equation in inhomogeneous media. Trudy Seismolog. Inst. **42**, p. 1 (1934).

74.* Sobolev, S. L.: Diffraction of waves by Riemann surfaces. Trudy Mat. Inst. Steklov **9** (1935).

75.* Sobolev, S. L.: A generalization of Kirchhoff's formula. Dokl. Akad. Nauk SSSR, Nov. Ser. **6** (1933).

76.* Sobolev, S. L.: A new method of solving Cauchy's problem for partial differential equations of normal hyperbolic type. Mat. Sb. **1** (43), 1 (1936).

77.* Sobolev, S. L.: Some Applications of Functional Analysis to Mathematical Physics. Novosibirsk: Sibirsk. Otdel. Akad. Nauk SSSR 1962.

78.* Some Methods and Algorithms for Interpreting Geophysical Data. Collection of articles edited by M. M. Lavrent'ev. Moscow: Nauka 1967.

79.* Sretenskiĭ, L. N.: On an inverse problem in potential theory. Izv. Akad. Nauk SSSR Ser. Mat. **5–6**, 551–570 (1938).

80.* Sretenskiĭ, L. N.: Newtonian Potential Theory. Moscow-Lenigrad: Gostehizdat 1964.

81.* Sretenskiĭ, L. N.: On the uniqueness of shape of an attracting body determined from the values of its exterior potential. Dokl. Akad. Nauk SSSR **99**, 1, 21–22 (1954).

82.* Tihonov, A. N.: On stability in inverse problems. Dokl. Akad. Nauk SSSR **39**, 5, 195–198 (1943).

83.* Tihonov, A. N.: On the solution of ill-posed problems and a regularization procedure. Dokl. Akad. Nauk SSSR **151**, 3, 501–504 (1963).

84. Wiechert, E., Zoeppritz, K.: Über Erdbebenwellen. Nachr. Königl. Gesellschaft Wiss. Göttingen **4**, 415–549 (1907).

Subject Index

Springer Tracts in Natural Philosophy

Vol. 1 Gundersen: Linearized Analysis of One-Dimensional Magnetohydrodynamic Flows. With 10 figures. X, 119 pages. 1964

Vol. 2 Walter: Differential- und Integral-Ungleichungen und ihre Anwendung bei Abschätzungs- und Eindeutigkeitsproblemen
Mit 18 Abbildungen. XIV, 269 Seiten. 1964

Vol. 3 Gaier: Konstruktive Methoden der konformen Abbildung
Mit 20 Abbildungen und 28 Tabellen. XIV, 294 Seiten. 1964

Vol. 4 Meinardus: Approximation von Funktionen und ihre numerische Behandlung
Mit 21 Abbildungen. VIII, 180 Seiten. 1964

Vol. 5 Coleman, Markovitz, Noll: Viscometric Flows of Non-Newtonian Fluids. Theory and Experiment. With 37 figures. XII, 130 pages. 1966

Vol. 6 Eckhaus: Studies in Non-Linear Stability Theory
With 12 figures. VIII, 117 pages. 1965

Vol. 7 Leimanis: The General Problem of the Motion of Coupled Rigid Bodies About a Fixed Point. With 66 figures. XVI, 337 pages. 1965

Vol. 8 Roseau: Vibrations non linéaires et théorie de la stabilité
Avec 7 figures. XII, 254 pages. 1966

Vol. 9 Brown: Magnetoelastic Interactions
With 13 figures. VIII, 155 pages. 1966

Vol. 10 Bunge: Foundations of Physics
With 5 figures. XII, 312 pages. 1967

Vol. 11 Lavrentiev: Some Improperly Posed Problems of Mathematical Physics
With 1 figure. VIII, 72 pages. 1967

Vol. 12 Kronmüller: Nachwirkung in Ferromagnetika
Mit 92 Abbildungen. XIV, 329 Seiten. 1968

Vol. 13 Meinardus: Approximation of Functions: Theory and Numerical Methods
With 21 figures. VIII, 198 pages. 1967

Vol. 14 Bell: The Physics of Large Deformation of Crystalline Solids
With 166 figures. X, 253 pages. 1968

Vol. 15 Buchholz: The Confluent Hypergeometric Function with Special Emphasis on its Applications. XVIII, 238 pages. 1969

Vol. 16 Slepian: Mathematical Foundations of Network Analysis
XI, 195 pages. 1968

Vol. 17 Gavalas: Nonlinear Differential Equations of Chemically Reacting Systems
With 10 figures. IX, 107 pages. 1968

Vol. 18 Marti: Introduction to the Theory of Bases
XII, 149 pages. 1969

Vol. 19 Knops/Payne: Uniqueness Theorems in Linear Elasticity
IX, 130 pages. 1971

Vol. 20 Edelen/Wilson: Relativity and the Question of Discretization in Astronomy
With 34 figures. XII, 186 pages. 1970

Vol. 21 McBride: Obtaining Generating Functions
XIII, 100 pages. 1971

Vol. 22 Day: The Thermodynamics of Simple Materials with Fading Memory
With 8 figures. X, 134 pages. 1972

Vol. 23 Stetter: Analysis of Discretization Methods for Ordinary Differential Equations
With 12 figures. XVI, 388 pages. 1973

Vol. 24 Strieder/Aris: Variational Methods Applied to Problems of Diffusion and Reaction. With 12 figures. IX, 109 pages. 1973

Vol. 25 Bohl: Monotonie: Lösbarkeit und Numerik bei Operatorgleichungen
Mit 9 Abbildungen. IX, 255 Seiten. 1974